Sequential Models of Mathematical Physics

Sequential Models of Mathematical Physics

Simon Serovajsky

CRC Press

Taylor & Francis Group

Boca Raton London New York

CRC Press is an imprint of the
Taylor & Francis Group, an **informa** business

CRC Press
Taylor & Francis Group
6000 Broken Sound Parkway NW, Suite 300
Boca Raton, FL 33487-2742

First issued in paperback 2020

ISBN-13: 978-1-138-60103-1 (hbk)
ISBN-13: 978-0-367-65665-2 (pbk)

Library of Congress Cataloging-in-Publication Data

Names: Serovajsky, Simon, author.
Title: Sequential models of mathematical physics / Simon Serovajsky
Description: Boca Raton, Florida : CRC Press, [2019] | Includes
bibliographical references and index.
Identifiers: LCCN 2018050822| ISBN 9781138601031 (hardback : alk.
paper) | ISBN 9780429470417 (ebook : alk. paper)
Subjects: LCSH: Mathematical physics. | Mathematical models. |
Mathematics--Methodology
Classification: LCC QC20 .S47 2019 | DDC 530.15--dc23
LC record available at https://lccn.loc.gov/2018050822

Visit the Taylor & Francis Web site at
http://www.taylorandfrancis.com

and the CRC Press Web site at
http://www.crcpress.com

To my wife Larissa

Contents

Preface

My teaching activities at the university began with a classical lecture course on the *"equation of mathematical physics"* for the students in the specialty *"mechanics"*. Given their professional interests, I began the course with practical applications. I wanted to show them that the equations of mathematical physics are of interest to them, first of all, because they are mathematical models of natural phenomena.

We have the clear scheme. The elementary volume of the region is chosen. Starting from the laws of physics, we estimate the difference between what enters and goes into this volume over a certain time interval. Then the passage to the limit is realized, when this volume is compressed into a point, and the length of the given time interval tends to zero. As a result, one or another equation of mathematical physics is obtained.... Many years later the question arose as to how justified this limiting transition is.

Norbert Wiener once wrote that the most productive for the development of the sciences are the areas relating to the "nobody's territory" between the different established sciences. Perhaps, this statement in the best degree characterizes the nature of the problem under consideration. On the one hand, the construction of a mathematical model of the physical process is really a translation of the specific laws of physics into the mathematical language. This is the business of physics. On the other hand, the passage to the limit is one of the most important mathematical procedures, which it is customary to treat with the utmost care. The justification of the limit transitions is a mathematical operation.

The majority of mathematicians, specialists in the equations of mathematical physics perceive a particular equation as a subject of research, given in a finished form. It is nice that this object has a certain physical sense. This can serve as a justification for the practical significance of the mathematical research. However, the main thing is the analysis itself, in particular, the existence of a solution in one sense or another under certain restrictions on the parameters of the problem, its uniqueness, smoothness, qualitative behavior, the properties of the dependence of the solution on parameters, etc. But the process of determination of the equation, according to many mathematicians, refers to physics....

Most physicists agree with this. However, they pay the greatest attention to the evaluation of physical factors that determine the corresponding balance relations in the chosen elementary volume. The justification of the

convergence is not relevant to their most important scientific interests. Therefore, they usually either simply formally pass to the limit without thinking
about the degree of validity of this procedure (the reasoning of the convergence is not theirs, but belongs to mathematicians) or they assume that the
functions under consideration are sufficiently smooth, as a result of which the
required limit actually exists. It would seem that the problem is solved. However, a new question arises: how do we know that the corresponding functions
have this degree of smoothness? The functions in question are precisely those
unknowns with respect to which the state equations are obtained. Further, it
would seem, everything is simple. The functional properties of the equations
of mathematical physics are established by means of a fairly well-developed
theory of certain equations. We use these results and obtain the desired properties of their solutions!

However, the last question arises. The above results of the theory of equations of mathematical physics have the following form: under certain constraints on the parameters of a given equation, its solution has the corresponding properties. Here it is assumed that the equation has already been
given. However, how can we establish the properties of the solution of the
equation in a situation, where the equation has not yet been obtained?

Thus, we have two obviously true statements. If a function has a proper
degree of smoothness, then it satisfies the corresponding equation. If the equation is given, then (under certain suppositions) its solution has a proper degree
of smoothness. However, these statements do not solve the above problem.
Over time, I had a desire to somehow understand this situation. Then there
was a special course, which I lectured for many years at the Mechanics and
Mathematics Faculty of the al-Farabi Kazakh National University. Then a
small book in Russian appeared (see [166]). Its main idea is described in the
paper [169]

The general structure of the book is given in the figure below (see Figure
1). In the first chapter we give a general formulation of the problem, based on
the concept of the classical solution of the problem of mathematical physics.
The subject of the following chapter is a generalized solution of the problem
of mathematical physics. Its connection with the classical solution is analyzed,
and its independence from the classical solution at the stage of determination
and practical implementation is shown. However, this concept has the same
drawback, and the question of the degree of validity of the passage to the limit
in the derivation of the mathematical model remains open.

These chapters constitute the first part of the book. Its second part, which
includes three chapters, is devoted to the general problem of justifying the
convergence. The third chapter describes different forms of sequence convergence. This is based on an estimate of the degree of proximity of an element
of a sequence with large enough number to an element called the limit. The
practical application of these definitions is difficult if the value of the limit
and even the fact of convergence remains unknown. In mathematical analysis,
at this step of analysis, the Cauchy convergence criterion is used, according

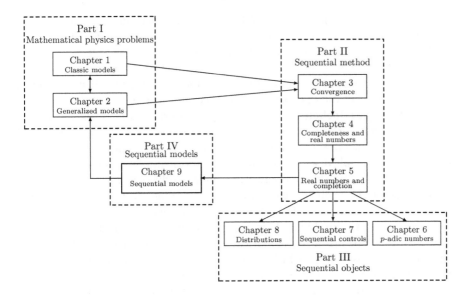

FIGURE 1: Structure of the book.

to which any fundamental sequence is convergent. The sequence is fundamental if its elements approach each other unboundedly. Checking this condition requires knowing only the elements of the sequence, which is acceptable in practice.

The fourth chapter begins with an analysis of examples of spaces, where the Cauchy criterion does not work, and hence the fundamental sequence may not converge. Thus, all spaces are divided into complete and incomplete, depending on whether the Cauchy criterion is true or not. We would like to prove the convergence for the incomplete spaces too. The way to overcome this difficulty is based on Cantor's definition of real numbers. There are equivalence classes of fundamental sequences of rational numbers. In this case, the divergent fundamental sequence of rational numbers determines an irrational number, and these different sequences can define the same real number. Such sequences are called equivalent. It is important that any fundamental sequences of real numbers converges, and any real number (the object of an extended set) can be arbitrarily accurately approximated by rational numbers (objects of the original set).

In the fifth chapter, we show that the objects defined in this way have all the properties that are assumed to be attributed to real numbers. Then the results obtained are extended to the general case. A theorem on the completion of a metric space is considered. By this result, any incomplete metric space can be extended to a complete space (its completion) so that any element of the completion can be arbitrarily closely approximated by the elements of the original set. This is the basis of the sequential method. Now we consider sequential objects that are classes of equivalent fundamental sequences.

The third part gives an overview of different sequential objects. In three of its chapters we consider p-adic numbers, sequential controls, and distributions. The last example is especially important, since a generalized approach in mathematical physics is based on the theory of distributions.

The final part includes a unique chapter. Here, a sequential method is used to determine the sequential form of the model. We consider classes of equivalent fundamental sequences of "approximate solutions" of the problem. Obtaining this result does not require any a priori supposition for the solution of the problem. It is characteristic that under additional restrictions on the parameters of the problem, the fundamental sequences under consideration converge to a function satisfying the integral identity underlying the generalized solution of the problem. This proves the justification of the generalized approach in problems of mathematical physics. The result established in the first section, according to which a sufficiently smooth generalized solution of the problem is its classical solution, makes it possible to substantiate the classical approach too. Thus, the problem initially posed is completely solved. Extremely important is the fact that the same algorithm is applied to the practical solving of the classical, generalized and sequential models of the system. Depending on the constraints on the parameters of the system, it can output to one or another form of the model.

We would like to simplify the technical calculations as much as possible without losing the essence of the problem. In this connection, we confine ourselves to considering the single extremely simple illustrative example. This is a stationary heat transfer phenomenon for the one-dimensional case. Note that in the process of answering one particular question, one has to turn to different questions of mathematical and mathematical physics, the theory of differential equations and computational mathematics, the theory of numbers and the theory of distributions, algebra and topology, mathematical analysis and the theory of optimal control. Thus, this book can serve as an illustration of the unity of mathematics, as well as the absence of clear boundaries between mathematics and physics. The book is recommended for a wide range of specialists and students of mathematical and physical specialties.

I express my gratitude to G. Bizhanova, Y. Gasimov and M. Ruzhansky, who got acquainted with the materials of the book and made a number of important remarks. I am deeply grateful to A. Teplov; without him, this book would not be finished. I am grateful to V. Shcherbak, who offered a cover for the Russian version of the book. I dedicate this book to my wife L. Ananyeva; without her understanding and support nothing would have happened.

Author

 **Simon Serovajsky** research interests include Optimal Control Theory, Mathematical Physics, Nonlinear Functional Analysis, Mathematical and Computational Modeling, History and Philosophy of Mathematics, and Foundations of Mathematics.

His scientific achievements include definition of the extended operator derivative, extended differentiability of the solution of the nonlinear infinite dimensional systems with respect to parameter without its Gateaux differentiability, extended differentiability of the inverse and implicit operators, necessary conditions of optimality for the nonlinear infinite dimensional control systems, definition of the weakened approximate solution of optimal control problems, definition of the sequential model of systems, and sequential extension of optimal control problems.

Education

He obtained his doctorate in Physical and Mathematical Sciences (Mathematical Physics), from Institute of Mathematics, National Academy of Sciences, Almaty (Kazakhstan) in 1994. He submitted a thesis on *Extended differentiability and optimal control for the nonlinear problems of the mathematical physics and* another thesis *on Variational inequalities for the optimal control problems.* He pursued his PhD in Differential Equations and Mathematical Physics from Institute of Mechanics and Mathematics, Academy of Sciences of Kazakhstan, Almaty (USSR) in 1983. He pursued his diploma in Applied Mathematics from Kazakh State University, Almaty (USSR) in 1976.

Professional Activities

He has been a Professor of the Differential Equations and Control Theory Department, al-Farabi Kazakh National University, Almaty (Kazakhstan) since 2011.

He was the Head of the Control Theory Department from 2009 to 2011. He was the Professor of the Calculus Mathematics Department from 1997 to 2009. He was the Head of the Applied Mathematics Department from 1995 to 1997. He was an Associate Professor from 1991 to 1995. He was an Associate Professor, Kazakh State University, Almaty (USSR) from 1985 to 1991. He was an Assistant Professor from 1983 to 1985. He was a Research Specialist from 1976 to 1983.

Simon Serovajsky has written the following books

1. Architecture of Mathematics. *Almaty, 21 Century*, 1998 (in Russian); *Second edition: Almaty, Print-S*, 2005.

2. Mathematical and Computer Models of Ecology (with N. Lyskovskaya and N. Popova). *Almaty, Kazakh University,* 1999 (in Russian); *Almaty, Print-S,* 2004 (with A. Karimov, in Kazakh).

3. Mathematical Modeling. *Almaty, Kazakh University,* 2000 (in Russian); *Almaty, Print-S,* 2004 (with A. Karimov, in Kazakh).

4. Counterexamples in the Optimal Control Theory. *Almaty, Kazakh University*, 2001 (in Russian); *Brill Academic Press. Netherlands, Utrecht-Boston*, 2004 (in English), http://www.allbookstores.com/author/ S_Y_Serovaiskii.html.

5. Introduction in Spectral Operators Theory. *Almaty, Alem,* 2003 (in Russian).

6. Sequential models of the Mathematical Physics. *Almaty, Alem,* 2004 (in Russian).

7. Optimization and Differentiation. Vol. 1. *Almaty, Print-S,* 2006 (in Russian).

8. Optimization and Differentiation. Vol. 2. *Almaty, Kazakh University,* 2009 (in Russian).

9. Practical Course of the Optimal Control Theory with Examples. 2011 *Almaty, Kazakh University* (in English).

10. Thinking about Mathematics and its History. *Almaty, Fast Polygraphy.* 2015 (in Russian).

11. Optimization and Differentiation. *CRS Press.* 2017. https://www. crcpress.com/Optimization-and-Differentiation/Serovajsky/p/book/978 1498750936 (in English).

12. History of Mathematics. Evolutions of mathematical ideas. In three volumes. *Moscow, URSS*, 2019 (in Russian).

Professor Simon Serovajsky, Ph.D.
Faculty of Mechanics and Mathematics
al-Farabi Kazakh National University
Almaty Republic of Kazakhstan

Part I

Mathematical physics problems

Well-known classical and generalized forms of solutions of mathematical physics problems are considered. The classic solution is a smooth enough function that satisfies the state equations with boundary conditions at each point of its domain. The generalized solution belongs to a Sobolev space and satisfies an integral equality. We consider relations between these notions, its determination from physical laws and methods of its practical finding. However, our general problem is the validity of the method of its definition.

Chapter 1

Classic models

We consider the classic method of obtaining mathematical models of physical phenomena. It consists of choosing an elementary volume, determining balance relations by means of physical laws, and passage to the limit as this volume shrinks to a point. This relates to the classic solution of mathematical physics problems. This is a smooth enough function that satisfies the state equations and the corresponding boundary conditions at the arbitrary point of its domain. This notion is used also for finding the approximate solution of the problem. However, we have some doubts about the correctness of the mathematical model definition by the classic approach.

1.1   Mathematical analysis of a physical phenomenon

Consider the application of mathematical methods for the analysis of a physical phenomenon. Its basis is a ***mathematical model***. This is a description of a system using mathematical concepts and language. The process of developing a mathematical model is called ***mathematical modeling***[1]. In reality, there exist three general stages of this analysis (see Figure 1.1):

   i) the definition of the mathematical model;

   ii) the analysis of the properties of the given model;

   iii) practical solving of this model.

We could add here the interpretation of the results of modeling and its practical application. However, these questions already go far beyond mathematics, and this does not apply to the discussed problem. Therefore, we confine ourselves to the three above-mentioned stages.

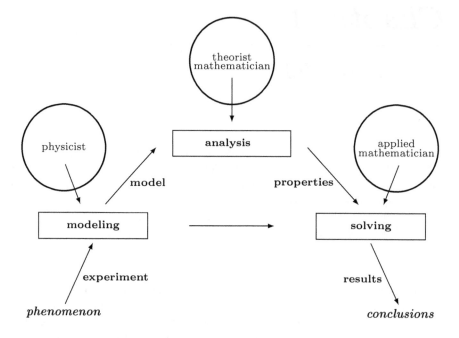

FIGURE 1.1: The scheme of the research.

The definition of the model is based directly on a natural experiment and physical laws. The analysis of the physical phenomenon is a work of the physicist. However, the obtained model is a real mathematical problem. Then the theorist mathematician analyses the concrete properties of this model. Particularly, one proves the existence of its solution under some suppositions with respect to parameters of the system, the uniqueness of this solution, its functional properties, the continuity of the solution with respect to system parameters, etc. However, we want not only to determine the general qualitative properties of the solution of the problem, but also to find this solution directly, if it is accurate, otherwise then at least approximately. The practical solving of the model is the business of the applied mathematician.

Unfortunately, such clear and seemingly quite natural delimitation of actions in the process of researching the problems of mathematical physics sometimes leads to undesirable consequences. The greatest troubles here should be expected at the intersection of individual stages, where the spheres of influence of different specialists are not so clearly determined.

At first, we would like to obtain the mathematical model of a physical phenomenon. There exist often the following standard steps of mathematical modeling (see Figure 1.2):

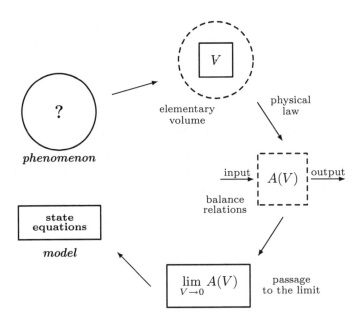

FIGURE 1.2: Definition of mathematical physics models.

i) The definition of an elementary volume from the considered set for the analyzing system.

ii) The determination of most important balance relations in this volume by means of physics laws that can be the laws of conservation of the main changing characteristics: energy, momentum, mass, charge, etc.

iii) Passage to the limit at these balance relations as this volume shrinks to a point.

The state equations are the result of the final step. There are equations with respect to general characteristics of the phenomenon, which are called *state functions*. The state equations with initial and boundary conditions give us as a rule the classic mathematical models. Try to realize this idea for a concrete example.

These actions are so customary. Therefore, researchers often do not pay proper attention to one extremely serious procedure. This is the passage to the limit that is one of the most important and at the same time the least obvious mathematical procedures. The problem of the convergence is fundamental for the mathematical analysis. However, in mathematical physics, the difficulties encountered in justification of passage to the limit in the construction of mathematical models are often not given due attention. Note that the insufficient validity of the procedure of the definition of state equations can call into question all subsequent actions on them, in particular, a qualitative and quantitative analysis of the system. Consider a concrete example.

1.2   Definition of a mathematical model

We would like to simplify all technical transformations. Therefore, we restrict ourselves to the consideration of an extremely simple and well-known example. Consider the stationary heat transfer phenomenon[2]. Let us have a non-homogeneous body with a length L under a source of the heat. Let this body be long and thin. Suppose that the characteristics of the system do not change with time. Then we can have a one-dimensional system. A detailed consideration of the determination of the corresponding mathematical model can be found in any classical course in mathematical physics. We give subsequent transformations because of the strong need to get a clear idea of the essence of our problem.

Choose an interval $[x, x + h]$ as an elementary volume for the one-dimensional case, where x is a point of the given set, and h is a length of the chosen interval. Determine the change of the quantity of the heat q there. We have the following equality

$$q(x) - q(x + h) = \int_x^{x+h} f(\xi)d\xi. \qquad (1.1)$$

The known function $f = f(x)$ characterizes the source of the heat here. Particularly, this determines the heat on the unit interval under this source. Therefore, the integral of the right-hand side of the last equality is the heat at the given interval of the body under the source (see Figure 1.3). Let the dimension of the heat flux be the same as the dimension of the spatial coordinate x. It is equal to the difference between the quantity of the heat at the beginning and at the end of the interval because of the equality (1.1).

The existence of the heat flux at a point x is the corollary of the difference between the temperature at this point and at the previous site. This is the basis of the heat conductivity phenomenon. By **Fourier's law**[3], the heat flux at a point x is proportional to the difference between the temperature at the beginning of the given interval and at its end. This is inversely proportional also to the length of this interval. Besides, the heat moves from the domain with a large temperature to the domain with a small temperature. Therefore, we obtain the formula (see Figure 1.4)

$$q(x) = -k(x)\frac{u(x) - u(x - h)}{h}, \qquad (1.2)$$

where u is the temperature, and k is the coefficient of the heat conductivity

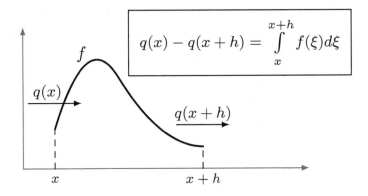

$$q(x) - q(x+h) = \int\limits_{x}^{x+h} f(\xi)d\xi$$

FIGURE 1.3: Change of the heat quantity.

of the body[4]. The temperature is our **state function**, because it describes the state of the considered system; and k is a known function. The length of the intervals at the equalities (1.1) and (1.2) can be different. However, we consider the easiest case of its equality.

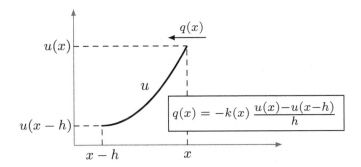

$$q(x) = -k(x)\frac{u(x)-u(x-h)}{h}$$

FIGURE 1.4: Fourier's law.

Suppose the function u is twice continuously differentiable at the given interval. After the passage to the limit at the equalities (1.1), (1.2) as $h \to 0$, we get the **non-homogeneous stationary one-dimensional heat equation**

$$\frac{d}{dx}\left[k(x)\frac{du(x)}{dx}\right] = f(x), \ x \in (0, L). \tag{1.3}$$

Remark 1.1 The considered equation has different physical interpretations. It can describe also the diffusion, the electrical conductivity, etc.[5]

Remark 1.2 We shall see that mathematical objects can have different mathematical interpretations. For example, the real numbers can be determined by fundamental sequences of rational numbers (see Chapter 4). However, there exist its interpretations as infinite decimals and cuts of rational numbers (see Chapter 5). We shall consider also different interpretations of p-adic numbers (see Chapter 6) and distributions (see Chapter 8).

Remark 1.3 The equation (1.3) is the simplest mathematical model of the heat transfer process. We assumed here the characteristics of the process unchanged with time, their variation solely with respect to one spatial variable, and also taking into account only the phenomenon of thermal conductivity and the action of thermal sources[6]. We consider this model solely for reasons of simplicity, since the discussed problems are already observed at the simplest situation. Naturally, in practice, as a rule, one applies with more meaningful (and, therefore, more difficult) models of the heat transfer process[6]. For them, the problems we are considering are also relevant. However, their analysis is required by a more difficult technique that does not directly concern the essence of the problems we are considering.

We suppose for simplicity that the temperature of the body on the boundary of the interval is equal to zero here. Then we have the boundary conditions

$$u(0) = 0, \ u(L) = 0. \tag{1.4}$$

Definition 1.1 *The boundary problem* (1.3), (1.4) *is the **classic mathematical model** of the considered phenomenon.*

Consider the properties of this model.

1.3 Classic solution of the system

We have the second order differential equation (1.3) with respect to the function u that satisfies also the homogeneous boundary conditions (1.4)[7]. Choose the natural functional class for this solution (see Figure 1.5). Determine the set $C_0^2[0, L]$ of twice continuously differentiable functions on the interval $[0, L]$ that are equal to the zero at the ends of this interval. We shall use sometimes the shorter denotation C_0^2.

Remark 1.4 One denotes often by $C_0^2[0, L]$ the set of all twice continuously differentiable functions on the interval $[0, L]$ such that these functions and their first derivatives are equal to the zero at the ends of the given interval.

Remark 1.5 The space $C_0^2[0, L]$ is Banach, that is complete linear normalized space. We shall consider this class of mathematical spaces in Chapter 4.

Determine the standard form of the solution for the boundary problem (1.3), (1.4).

Definition 1.2 *The function u from the set $C_0^2[0, L]$ is called the **classic solution** of the boundary problem* (1.3), (1.4), *if it satisfies the equality* (1.3) *for all points x of the open interval* $(0, L)$.

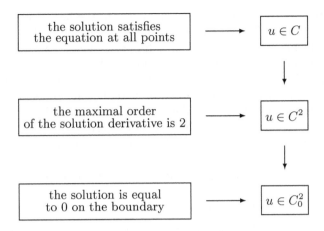

FIGURE 1.5: Choice of the functional class for the classic solution.

Note that the definition of the mathematical model is associated with the notion of the classic solution of the considered mathematical physics problem because the possibility of passage to the limit at the balance relations requires the functional properties of the classic solution for the considered state function.

We have also an interest in the qualitative and quantitative analysis of this mathematical model. The qualitative analysis is, at first, the proof of solvability of the system. It is necessary to prove that the boundary problem (1.3), (1.4) has a classic solution under some properties of the known functions f and k. We could determine also the uniqueness of the solution, its continuous dependence from the parameters of the systems, etc. These results can be obtained by using the differential equations theory, in reality. However, we do not determine these properties as yet.

Remark 1.6 An existence theorem of the classic solution for the considered boundary problem will be considered in Chapter 9.

Consider now the final step of the analysis. This is practical solving of the boundary problem.

1.4 Approximate solution of the system

The best result of the analysis of the considered problem could be obtaining the analytic formula of its solution. This is the direct dependence of the temperature u from the space coordinate x for all values of all parameters of the system, namely the length of the body L, the heat source f, and the heat conductivity k. Unfortunately, the analytic solution of the mathematical physics problems can be obtained only for easy enough partial cases[8]. However, we can find the approximate solution of the problems for the general case. There exist many approximate methods of solving for the problem (1.3), (1.4). We choose the ***finite difference method*** because of its high enough effectiveness and simplicity. It uses the division of the given domain of the state function by parts and the approximation of its derivatives by the corresponding differences[9]. Determine, at first, the easiest formulas of the ***approximate differentiation***.

Consider a function $y = y(x)$. Suppose it is differentiable. Then we have the equality

$$y(x + h) = y(x) + y'(x)h + o(h),$$

by the Taylor's formula, where $o(h)/h \to 0$ as $h \to 0$. If the value h is small enough, we have the approximate equality

$$y(x + h) \approx y(x) + y'(x)h.$$

Therefore, we can find

$$y'(x) \approx \frac{y(x + h) - y(x)}{h}.$$

This formula of the approximate differentiation is called the ***forward difference***. Its exactness is determined by the value of the step h.

Analogically, from the formula with negative step

$$y(x - h) = y(x) - y'(x)h + o(h)$$

it follows the approximate equality

$$y(x - h) = y(x) - y'(x)h.$$

Then we get

$$y'(x) \approx \frac{y(x) - y(x - h)}{h}.$$

This formula of the approximate differentiation is called the ***back difference***.

Using the formulas of approximate differentiation, determine the algorithm of finding an approximate solution of the boundary problem. Divide the given interval $(0, L)$ by M equal parts. Choose the step $h = L/M$ and the points $x_i = ih$, $i = 0, 1, ..., M$. Determine the standard difference operators on the Euclidean spaces

$$\delta_{\bar{x}} : \mathbb{R}^{M+1} \to \mathbb{R}^M, \quad \delta_x : \mathbb{R}^{M+1} \to \mathbb{R}^M$$

by the equalities

$$\delta_{\bar{x}} y_i = \frac{y_i - y_{i-1}}{h}, \quad i = 1, ..., M,$$

$$\delta_x y_i = \frac{y_{i+1} - y_i}{h}, \; i = 0, ..., M-1,$$

where $y_i = y(x_i)$.

Consider the equality (1.1) at the arbitrary point x_i. We have

$$\delta_x q_i = \frac{q_i - q_{i+1}}{h} = \frac{1}{h} \int_{x_i}^{x_{i+1}} f(\xi)d\xi, \tag{1.5}$$

where $q_i = q(x_i)$. Analogically, from the equality (1.2) it follows that

$$q_i = -k_i \frac{q_i - q_{i-1}}{h} = -k_i \delta_{\bar{x}} u_i, \tag{1.6}$$

where $u_i = u(x_i)$, $k_i = k(x_i)$. Put the value q_i from the equality (1.6) to (1.5). We obtain

$$\delta_x (k_i \delta_{\bar{x}} u_i) = f_i, \; i = 1, ..., M-1, \tag{1.7}$$

where

$$f_i = \int_{x_i}^{x_{i+1}} f(\xi)d\xi.$$

We use the difference equation (1.7) as the approximation of the differential equation (1.3). Add also the boundary conditions

$$u_0 = 0, \; u_M = 0. \tag{1.8}$$

We have the system of the linear algebraic equations (1.7), (1.8) with triangle matrix. This problem can be solved by means of standard methods[10]. Therefore, we find all values u_i, namely the **grid function**.

Note that our solution is the function of the continuous argument. Then we use the **linear interpolation**[11] of the grid function (see Figure 1.6)

$$u_h(x) = u_i + x\delta_x u_i, \; x \in (x_i, x_{i+1}), \; i = 1, ..., M-1.$$

Besides, the function u_h is equal to zero on the boundary of our domain.

We would like to use u_h for a small enough value h as an approximate solution of the considered problem. Therefore, it is necessary to prove the convergence $u_h \to u$ as $h \to 0$. This is the convergence in the class of the twice continuously differentiable functions, because we would like to determine the classical solution of our boundary problem as the limit. However, the classic solution is an element of this functional space. This is the substantiation of the numerical method and the basis of the practical solution of the problem.

Remark 1.7 We shall return to the convergence of the finite difference method in Chapter 9.

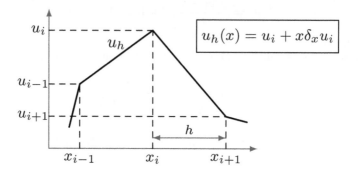

FIGURE 1.6: Linear interpolation of the grid function.

Note that the properties of the classical solution are used for all three steps of the research (see Figure 1.7). Indeed, the functional class $C_0^2[0, L]$ is on the base of the notion of the classical solution. We use it also for passage to the limit at the balance relations. Besides, it was be used for the determination and the substantiation of the method of finding the approximate solution of the problem. This indicates the existence of a serious connection between the different stages of solving of mathematical physics problems.

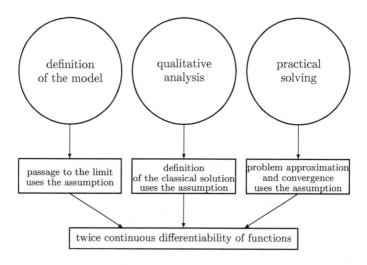

FIGURE 1.7: Assumption about the state function for the classical method.

Remark 1.8 We have to make sure that this connection is extremely deep. Moreover, the separation of the research process into stages is conditional.

Now we consider the central problem of this book. This is the validity of the process of obtaining this mathematical model.

1.5 Validity of the classic method

We obtained our mathematical model by means of the passage to the limit at the balance relations as the length of the elementary interval tends to zero. As is known, the limit is the general notion of mathematical analysis. The convergence is one of the most important and difficult mathematical operations, in the realization of which it is necessary to have the maximum caution. This operation is surprisingly effective, but remains mysterious, because of its direct connection with infinite procedures. Therefore, one certainly requires a rigorous justification. The justification of the convergence is often the main technical difficulty at the theory of partial differential equations, computational mathematics, the theory of optimal control, and other branches of mathematics.

We would like to pass to the limit at the equalities (1.1), (1.2). However, why does this limit exist? Our analysis is correct, if the state function u is twice continuously differentiable. Indeed, we used this hypothesis before. Therefore, we have the next question. Why does function satisfy this property?

Maybe, we could use physical reasons. Perhaps, the temperature of the body must be a smooth function. However, our body can be non-homogeneous by our suppositions. Moreover, the heat source can have different properties. We do not know if the temperature is twice continuously differentiable for these cases. Besides, our object of analysis is the mathematical model, but not a physical body. We do not know if the properties of the phenomenon and its model are the same. Therefore, we need to use mathematical reasons only.

Maybe, we could prove the desired property of the considered function directly. Indeed, we could use the results of the differential equations theory[12]. In reality, the solution of the given boundary problem is twice continuously differentiable under some assumptions with respect to the known functions k and f. This is the result of the second stage of the analysis. This is obtained by the proof of the existence of the classical solution of the boundary problem. Unfortunately, this result can only be obtained after the determination of the mathematical model. We cannot possibly analyze the equation before its definition. This is a strange enough result. The determination of the model uses the properties of its state function. However, it is necessary to have the model for the analysis of this state function (see Figure 1.8).

This considered difficulty often remains without due attention. This is explained, apparently, by the fact that physicists, as well as chemists, biologists, economists, etc. are engaged generally in immediate modeling of natural processes. Indeed, the main problem for them is to identify those reasons that

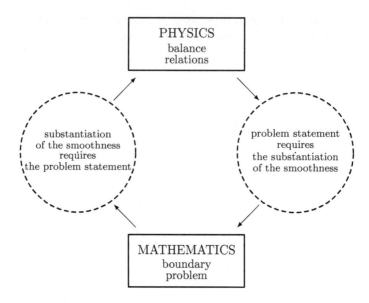

FIGURE 1.8: Relations between obtaining the mathematical model and properties of state functions.

predetermined the considered events. They establish the qualitative and quantitative impact of each of the considered factors on the result. In our case, it is important for the physicist that the investigated body is sufficiently long and thin, the characteristics of the phenomenon do not change, heat transfer is due solely to heat conduction; besides these, the convection, the radiation, the heat exchange with the environment, the presence of chemical reactions, and other possible causes of heat transfer, are not important here. The physicist tries to understand the studied phenomenon and often does not think about the technical mathematical difficulties. Indeed, the justification of mathematical procedures is the business of the professional mathematicians.

On the other hand, the professional mathematicians usually consider the equations as an immediate given object of research. They, as a rule, believe that the derivation of equations is the work of the physicists. Indeed, we have the statement of the mathematical problem, for example, a boundary problem for a differential equation. The mathematicians need to prove the existence of the given problem and find the algorithm of its solving. An equation is a real, existing object for the theoretical mathematicians, as natural as an electrical circuit for the physicists, a chemical reagent for chemists, and a living organism for a biologists. The equations are the serious objects of analysis for the mathematicians even if they do not have any physical sense.

We could suppose that this problem is not very important. Indeed, the absolute majority of physicists and mathematicians are not very interested, apparently, in its solution. Maybe this is really so. However, if we have the right to carry out formally one of the stages of the analysis of the problem without a rigorous mathematical justification, then we must also apply it to its other stages. Is it necessary to prove the existence of a solution to a problem at all, if already from the physical considerations it follows, for example, that the body at any point possesses a certain temperature? Therefore, the required solution must certainly exist, and there is no problem here. Many researchers of practice reason this way.... Indeed, we can write down and program all the necessary formulas, and the computer will definitely produce some result.... However, how do we know that this is indeed the solution of the given equations? Of course, we can compare the results with an experiment. However, we are not sure that our model describes the considered phenomenon exactly. Besides, we do not know exactly the parameters of the system, for example, the functions k and f in our example. Therefore, it is not clear how we can use the results of the experiment for proving the existence of the solution. We can use results of computing. However, can we guarantee that we found the approximate solution of the problem without the proof of the convergence of the numerical algorithm? If the mathematical rigor is needed at the stage of qualitative and quantitative analysis of the system, then, apparently, it is needed too in the construction of a mathematical model. It is better to find an approximate solution of the problem without proving its solvability and convergence of the algorithm, than not to obtain any result. However, this is the only evidence of our weakness.

Thus, we apparently have good reasons for doubting the correctness of the derivation of mathematical models by available means. This exceptionally serious circumstance can cast doubt on the validity of the classical approach in mathematical physics. Therefore, there is a need to find another way for constructing mathematical models that do not possess these disadvantages. In particular, one can try to pass from the concept of the classical solution of the problem to a generalized solution. As we know, the most important directions in the development of mathematical physics have been connected for a long time already with the generalized, and not with the classical approach. Therefore, we suppose that this could be a basis of the correct constructing of mathematical models.

1.6 Conclusions

1. Mathematical analysis of physical phenomenon contains the definition of the mathematical model, the determination of the properties of state functions and finding of the solution of the problem.

2. All stages of this analysis are based on the classic solution of the problem.

3. The classic solution of the problem is a twice differentiable function that satisfies the state equation at the arbitrary point and the boundary conditions.

4. The classic mathematical model is the result of passage to the limit at the balance relations that are the corollary of physical laws.

5. The limit exists if the state function is twice differentiable.

6. The necessary properties of the state function can be obtained by means of the differential equations theory after the determination of the mathematical model.

7. We cannot have any information about the state function before the determination of the mathematical model.

8. The classic method of the determination of the mathematical model is not substantiated.

9. It is necessary to find another method of analysis.

The modern theory of equations of mathematical physics often uses the concept of a generalized solution of the problem in place of the classical one. Therefore, we shall try to substantiate the process of mathematical modeling by means of a generalized approach.

Notes

^{1}*Pythagoras* taught that the knowledge of the world is reduced to the knowledge of numbers that govern the world. In a certain sense, this idea underlies mathematical modeling. The general methods of mathematical modeling are considered in [19], [44], [59], [161], [202].

^{2}The problems of heat physics are considered, for example, in [16], [17], [32], [52], [72], [118], [134], [148].

^{3}Fourier's law was formulated by *Jean-Baptiste Joseph Fourier* in 1807. The multi-dimensional form of the Fourier law is $q = -k\nabla u$, where q is the heat flux density, k is the material's conductivity, and ∇u is the temperature gradient.

^{4}The Fourier's law has the diffusive analogue. This is **Fick's law** according to which the diffusive is proportional to the difference between the concentration at the begin of the given interval and at its end, and this is inversely proportional also to the length of this interval.

^{5}Heat transfer and diffusion (as well as charge transfer) belong to the class of transfer processes. The state of the system is described here by some characteristic (temperature, concentration, charge density, etc.), which is unevenly distributed over a given region. Further, for some reason, there is a flow of the relevant substance (heat, mass, charge, etc.) from a part of the region where there is an excess of it into the area where there is a defect. Thus, the system tends to some equilibrium state. Different non-physical interpretations of transfer processes are considered in [161].

^{6}The classic non-stationary **heat equation** for a homogeneous body has the form
$$\frac{\partial u}{\partial t} = a^2 \triangle u + f,$$
where a is the thermal diffusivity, and $\triangle$ is the Laplace operator, that is the sum of second spacial derivatives. This equation was be determined by *Jean-Baptiste Joseph Fourier*. Extensions of the considered equation are considered in in [16], [17], [32], [52], [72].

^{7}The theory of boundary problems for the second order differential equations is described, for example, in [36], [65], [80], [111], [138]. The multi-dimension analogue of the considered equation is the **elliptic equation**, and its non-stationary analogue is the parabolic equation. There are **partial**

differential equations or ***equations of mathematical physics***; see, for
example, [4], [20], [53], [101], [102], [103], [114], [115], [126], [128], [137], [184].

[8] Suppose, for example, the coefficient k is constant. Integrating the equality (1.3) two times and using the boundary conditions (1.4), we get the exact
formula of the solution of the considered boundary problem

$$u(x) = -\frac{x}{kL} \int_0^L \int_0^\xi f(\eta)d\eta d\xi + \frac{1}{k} \int_0^x \int_0^\xi f(\eta)d\eta d\xi.$$

[9]The finite difference method for solving differential equations is described,
for example, in [61], [130], [143], [150], [176], [180], [190].

[10]The system of the linear algebraic equations with triangle matrix can
be solved by ***tridiagonal matrix algorithm*** (Thomas algorithm); see, for
example, [15], [41], [139], [150]. This method was be proposed by *Israil Gelfand*
and *Oleg Likutsyevsky* in 1952.

[11]The linear interpolation was known to *Ptolemy*. The theory of interpolation is considered, for example, in [15], [96].

[12]The theory of ordinary differential equations is described, for example, in
[13], [36], [65], [80], [138], [206].

Chapter 2

Generalized models

We tried to obtain the mathematical model of the considered phenomenon by means of the classic approach. However, we had serious difficulty because we do not know how we can pass to the limit at the balance relations without a priori properties of the state function. These properties correspond to the classic solution of the mathematical physics problems. Now we change the classic model by the generalized one. This is based on the generalized solution of the mathematical physics problems (Section 2.1) that has many advantages over the classic one. We determine relations between classic and generalized solutions. Then the direct physical sense of the generalized solution is discussed (Section 2.2). We describe also the approximation method for finding the generalized solution of the problem without using its classic solution (see Section 2.3 and Section 2.4). The final step here is the justification of the problem of the determination of the generalized model (Section 2.5).

2.1   Generalized solution of the problem

It is well known that the proof of the existence of the classic solution and the convergence of the standard numerical algorithm to this solution are difficult enough problems. These require also strong enough limitations with respect to parameters of the systems. We talk, of course, about the difficulties of the classic approach to mathematical physics problems, in principle, not about the concrete example. These difficulties are largely overcome by using the concept of a generalized solution of the problem, based on the theory of distributions[1]. In this regard, we can hope for successfully overcoming the previously mentioned difficulties with a generalized approach.

Let us return to the easiest mathematical model of the stationary heat transfer phenomenon (see Chapter 1). We have the boundary problem

$$\frac{d}{dx}\left[k(x)\frac{du(x)}{dx}\right] = f(x), \ x \in (0, L), \tag{2.1}$$

$$u(0) = 0, \ u(L) = 0, \tag{2.2}$$

where $u = u(x)$ is the temperature of the body at the point x, $k = k(x)$ is the coefficient of the heat conductivity of the body, $f = f(x)$ characterizes the source of the heat, and L is the length of the body.

One knows that the classic solution of this problem is a twice differentiable function on the interval $[0, L]$ with zero values at the ends of this interval that is a point of the space $C_0^2[0, L]$; besides, it satisfies the state equation (2.1) at each point (see Chapter 1). Determine another form of the solution. It will be a function of the Sobolev space $H_0^1(0, L)$ of all square Lebesgue integrable functions on the interval $(0, L)$ with its first derivatives such that its values on the boundary of this interval are equal to zero. We shall use sometimes the short denotation H_0^1. The choice of this space in our situation is clarified by Figure 2.1.

Remark 2.1 The space $H_0^1(0, L)$ is the Hilbert space that is a complete linear space with scalar product, see Chapter 4.

Remark 2.2 We use the generalized derivatives here. Its definition will be given later (see Section 2.3). Note that the strict general definition of the generalized derivatives is determined by means of the distributions theory (see Chapter 8). The exact definition of Sobolev spaces (not only considered above) will be given in that chapter too.

Definition 2.1 *The **generalized solution** of the problem* (2.1), (2.2) *is an element of the space* $H_0^1(0, L)$ *such that the following integral equality holds*

$$-\int_0^L k(x)\frac{d\lambda(x)}{dx}\frac{du(x)}{dx}dx = \int_0^L \lambda(x)f(x)dx \ \forall \lambda \in H_0^1(0, L). \tag{2.3}$$

Remark 2.3 The existence and uniqueness theorem for the generalized solution of the considered boundary value problem for the stationary heat transfer equation will be given in Chapter 9.

Determine the relations between the classic and the generalized solutions.

Theorem 2.1 *Each classic solution of the problem* (2.1), (2.2) *is its generalized solution.*

Proof. Let the function u be a classic solution of the boundary problem (2.1), (2.2). Therefore, it belongs to the considered Sobolev space. Multiply

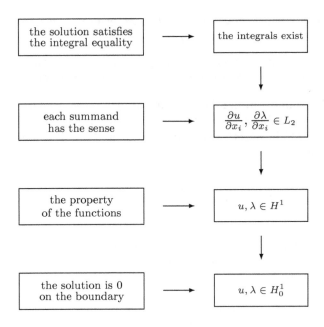

FIGURE 2.1: Choice of the functional class for the generalized solution.

the equality (2.1) by an arbitrary element λ of this Sobolev space. Integrating in x, we obtain

$$\int_0^L \frac{d}{dx}\left[k(x)\frac{du(x)}{dx}\right]\lambda(x)dx = \int_0^L f(x)\lambda(x)dx. \qquad (2.4)$$

After integration by parts at the left hand-side of this equality with using the equality of the function λ to zero on the boundary of the given interval, we have the equality (2.3). $\square$

Determine the inverse result.

Theorem 2.2 *The twice continuously differentiable generalized solution of the problem* (2.1), (2.2) *is its classic solution.*

We prove this result by using the following **fundamental lemma of the calculus of variations**[2].

Lemma 2.1 *If the continuous function f on an interval (a, b) satisfies the equality*

$$\int_a^b f(x)h(x)dx = 0$$

for all smooth enough functions h with zero values on the boundary of the given interval, then f is identically zero.

Remark 2.4 We could use here the definition of zero element of the adjoint space (see Chapter 3) instead of the following fundamental lemma of the calculus of variations.

Proof of Theorem 2.2. Let the function u be a twice continuously differentiable generalized solution of the boundary problem (2.1), (2.2). Integrate the equality (2.3) by parts. We obtain the equality (2.3). It can be transformed to the equality

$$\int_0^L \left\{ \frac{d}{dx}\left[k(x)\frac{du(x)}{dx}\right] - f(x) \right\}\lambda(x)dx = 0.$$

Therefore, the equality (2.1) is true at the arbitrary point x because of Lemma 2.1 and the arbitrariness of the function λ. We have also the boundary conditions (2.2) by the definition of the Sobolev space. □

Of course, the function of the Sobolev space cannot be twice differentiable. Therefore, the generalized solution of the problem is not necessarily its classic solution. However, the twice-differentiable generalized solution is the classic solution of the problem. It is clear that a non-smooth generalized solution is not the classic solution. The relation between the classic and the generalized solutions is imaged on the Figure 2.2.

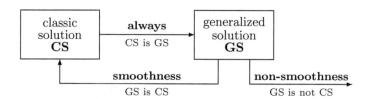

FIGURE 2.2: Relation between classic and generalized solutions.

Remark 2.5 The generalized solution of the problem is an element of the larger set than the classic one. Therefore, the uniqueness of the classic solution follows from the uniqueness of the generalized solution. However, the generalized solution can be non-unique for the case of the uniqueness of the classic solution, because the additional generalized solution can be non-smooth. The quantity of the solutions is greater the wider the set in which these solutions are determined. For example, the algebraic equation $x^3 - x = 0$ has unique solution on the set of positive numbers, two solutions on the set of non-negative numbers, and three solutions on the set of real numbers (see Table 2.1).

TABLE 2.1: Domains of the function
$f(x) = x^3 - x$ and the sets of the equation $x^3 - x = 0$

domains of the function object	set of the solutions
$X_1 = \{x \mid x > 0\}$	$X_1^s = \{1\}$
$X_2 = \{x \mid x \geq 0\}$	$X_2^s = \{1, 0\}$
$X_3 = \mathbb{R}$	$X_3^s = \{1, 0, -1\}$

Remark 2.6 We shall determine also the sequential state of the considered system in Chapter 9. The relation between the sequential and generalized states is an analogue of the generalized and classic states.

Remark 2.7 The generalized solution can be determined for the easiest algebraic equation (see Table 2.2). Particularly the *additive equation* $a + x = f$ can be considered on the set of natural number $\mathbb{N}$. The parameters a, f, and the solution x are chosen from the set $\mathbb{N}$ here. This equation can be solvable (for example, $2 + x = 3$) and insolvable (for example, $3 + x = 2$). However, we can declare that last equation has the generalized solution $x = -1$, which is the element of the larger set of integer numbers $\mathbb{Z}$. Analogically, the *multiplicative equation* $a \cdot x = f$ can be considered on the set of integer number $\mathbb{Z}$. This equation can have the classic solution, for example, $2 \cdot x = 6$. However, the equation $6 \cdot x = 2$ is insolvable in the classic sense. But there exists the value $x = 1/3$ from the larger set $\mathbb{Q}$ of rational numbers that can be interpreted as the generalized solution of the considered equation. The analogical properties are true for the *square equation* $ax^2 + bx + c = 0$ and general *algebraic equation* $a_n x^n + a_{n-1} x^{n-1} + ... + a_1 x + a_0 = 0$ that are considered initially on the set of rational number $\mathbb{Q}$. Sometimes it has the classic solution, for example, the equation $x^2 = 1$. However, this solution can be absent for other parameters. For this situation we can determine its generalized solution. Particularly, the equation $x^2 = 2$ has two generalized solutions $x_1 = \sqrt{2}$ and $x_2 = -\sqrt{2}$ from the larger set $\mathbb{R}$ of real numbers. However, sometimes this equation does not have any solution from this set. Therefore, it is necessary to extend the class of the solutions. For example, the equation $x^2 = -1$ has two generalized solutions $x_1 = i$ and $x_2 = -i$ that are the elements of the larger set of complex numbers. The last result shows that for the same problem, different types of generalized solutions can be defined depending on the choice of the class of objects on which these solutions are determined.

Actually, the introduction of negative, fractional, etc. numbers has historically been stimulated mainly by the desire to obtain unconditionally solvable additive, multiplicative, etc. equations. These questions also lead to abstract algebra, in particular, to the theory of *groups* (see Chapter 5). For example, the solvability of the additive equation $a + x = f$ and multiplicative equation $a \cdot x = f$ is the corollary of the existence of the inverse element for the number a. The element is inverse to the given element, if the result of considered operation (addition or multiplication) between these elements gives the unit. This is such an element that the result of its operation with an arbitrary element of the considered set does not change this element. Particularly, the unit with respect to the addition is the number 0, and the unit with respect to the multiplication is the number 1. Then the number $-a$ is the inverse element to the number a with respect to the addition; and the fraction $1/a$ is its inverse element with respect to the multiplication. The unconditional existence of the inverse element is the property of group. The set $\mathbb{N}$ is not the group with respect to the addition. Then the additive equation can be insolvable on the set $\mathbb{N}$. However, the set $\mathbb{Z}$ of integer numbers with addition is the group. Therefore, the additive equation is solvable everywhere here. Analogically, the set $\mathbb{Z}$ with multiplication is not the group, and the set $\mathbb{Q}$ (more exact, the set of non-zero rational numbers) is the group. This determines the properties of the multiplicative equation.

TABLE 2.2: Generalized solutions

class of problems	concrete object	given set	classic solution	extended set	generalized solution
additive equation	$2 + x = 3$	$\mathbb{N}$	$x = 1$	$\mathbb{Z}$	$x = 1$
additive equation	$3 + x = 2$	$\mathbb{N}$	absence	$\mathbb{Z}$	$x = -1$
multiplicative equation	$2 \cdot x = 6$	$\mathbb{Z}$	$x = 3$	$\mathbb{Q}$	$x = 3$
multiplicative equation	$6 \cdot x = 2$	$\mathbb{Z}$	absence	$\mathbb{Q}$	$x = 1/3$
square equation	$x^2 = 1$	$\mathbb{Q}$	$x_1 = 1$ $x_2 = -1$	$\mathbb{R}$	$x_1 = 1$ $x_2 = -1$
square equation	$x^2 = 2$	$\mathbb{Q}$	absence	$\mathbb{R}$	$x_1 = \sqrt{2}$ $x_2 = -\sqrt{2}$
square equation	$x^2 = -1$	$\mathbb{Q}$	absence	$\mathbb{C}$	$x_1 = i$ $x_2 = -i$
minimization problems	$f(x) = (x - 1)^2$	$\{x \mid x > 0\}$	$x = 1$	$\{x \mid x \geq 0\}$	$x = 1$
minimization problems	$f(x) = x^2$	$\{x \mid x > 0\}$	absence	$\{x \mid x \geq 0\}$	$x = 0$

Remark 2.8 We can determine also the generalized solution for the extremum problems. Consider, for example, the easiest problem of minimization of a function $f = f(x)$ on a set U (see Table 2.2). Sometimes this problem is solvable. We can call its solution by a classic one. For example, the minimization problem for the function $f(x) = (x - 1)^2$ on the set of positive numbers has the solution $x = 1$. However, the minimization problem for the function $f(x) = x^2$ is insolvable here. This problem has the solution $x = 0$ on the extended set of non-negative numbers that can be interpreted as the generalized solution of the initial extremum problem. We shall consider the solvability properties of the extremum problem in Chapter 7.

The generalized solution has weaker functional properties than the classic one. Therefore, the proof of the solvability of the problem and the convergence of the numerical method for the generalized method are easier than the classic one. Besides, it uses weaker suppositions with respect to parameters of the system. These results and the equality between the smooth generalized solution and the classic one explain the popularity of the generalized method in mathematical physics.

Now we can hope to explore the justification of the mathematical models determination by the generalized method. However, we have a serious question. The generalized solution is determined for the boundary problem (2.1), (2.2). Therefore, the generalized solution seems to be the corollary of the classic one. The generalized approach could solve the problem of the justification of the mathematical model, if the integral equality (2.3) can be obtained directly from the physical law without using the boundary problem (2.1), (2.2). It is necessary to prove the direct physical sense of the equality (2.3). Try to obtain this result.

2.2   Determination of the generalized model

We would like to prove that the equality (2.3) has the direct physical sense. Consider again the equalities

$$q(x) - q(x+h) = \int_{x}^{x+h} f(\xi)d\xi, \tag{2.5}$$

$$q(x) = -k(x)\frac{u(x) - u(x-h)}{h} \tag{2.6}$$

that are the basis of the classic model (see Chapter 1).

Multiply the equality (2.5) by an arbitrary smooth enough function λ with zero values on the boundary, and integrate the result. Dividing by the interval length h, we get

$$\int_{0}^{L} \frac{\lambda(x) - \lambda(x-h)}{h}q(x)dx + \frac{1}{h}\int_{0}^{h}\lambda(x-h)q(x)dx -$$

$$\frac{1}{h}\int_{L}^{L+h}\lambda(x-h)q(x)dx = \int_{0}^{L}\lambda(x)\frac{1}{h}\int_{x}^{x+h}f(\xi)d\xi dx \ \forall \lambda \in H_0^1.$$

Our next transformations use the ***mean value theorem for integrals*** that is the well-known result of mathematical analysis[3] (see Figure 2.3).

Theorem 2.3 *Let the function f be continuous on the interval $[a, b]$. Then there exists a point c from this interval such that*

$$\int_{a}^{b} f(x)dx = f(c)(b-a).$$

Pass to the limit formally here using Theorem 2.3 and the equality to zero of the function λ at the ends of the given interval. We obtain the integral equality (2.3).

Thus, the equality (2.3) can be obtained as the corollary of the balance relations (2.5), (2.6). Therefore, it has the direct physical sense. Hence, we can interpret this equality as the special form of the mathematical model of the considered physical phenomenon. Now we can correct our definitions.

Definition 2.2 *The boundary problem* (2.1), (2.2) *is called the **classic model** of the considered process; and the integral equality* (2.3) *is called its **generalized model**.*

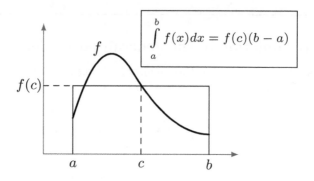

FIGURE 2.3: Equality of the areas of the curvilinear trapezoid and the rectangle.

Despite the final result, we still doubt the interpretation of the equality (2.3) as a model because of the arbitrariness of the function λ. Indeed, what sense has the mathematical model that depends on the arbitrary function? In reality, this is not a very serious problem. The classic model describes the state of the considered system directly. However, another case is possible for predicting the response of the system to each exterior influence. We can interpret the function λ as an exterior signal (see Figure 2.4). Moreover, the interior structure of the system can often be unknown. We can observe the properties of the system only by its response to the exterior signal. It conforms to the well-known **black box** notion[4].

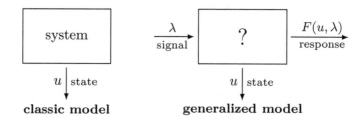

FIGURE 2.4: Classic and generalized models.

Remember the problems of the microcosm described by the quantum mechanics laws[5]? Any experimental fact here can be obtained exclusively through active intervention in the process. Of course, each experiment is, in principle, a result of the interaction between the considered system and the instrumentation. However, unlike the macrocosm in the quantum mechanical system, it is impossible to make measurements without introducing serious perturbations into the process under investigation. We cannot find out, in principle, where, for example, an electron is located in itself. The desired information can be obtained only by acting on it with something, for example, a photon. However, because of this experiment, the electron acquires somewhat different properties, in comparison with those that it originally had. The original properties of the object fully remain unclear; this is connected with the **Heisenberg uncertainty relation**. The classical form of the model is based on the direct description of the structure of the object, and the generalized one describes the response of the system to any external influence. Thus, the nature of the microcosm itself is such that the generalized approach is perfectly associated with the physics of the process.

We determined the function u of the Sobolev space that satisfies the equality (2.3) as the generalized solution of the boundary problem (2.1), (2.2). However, if it is non-smooth, then the equation (2.1) does not have any natural sense. Therefore, this boundary problem can be interpreted as the short form of the denotation of the problem (2.3) only. It is not the corollary of the physical law in this case, because we cannot admit any possibilities to pass to the limit at the balance relations. However, if the solution of the problem (2.3) is smooth enough, then we can deduce the boundary problem (2.1), (2.2) from the equality (2.3). Therefore, the classic model is, in reality, secondary with respect to the generalized one. Thus, it will be better to change our terminology, because it is not correct to use the denotation generalized solution of the boundary problem, if the equality (2.3) can be true for the case of non-applicability of this boundary problem. We give the following definition (see Figure 2.5).

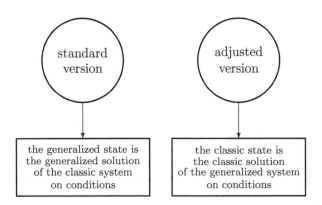

FIGURE 2.5: Relations between classic and generalized methods.

Definition 2.3 *The function of the Sobolev space $H_0^1(0, L)$ is called the **generalized state** of the considered system, if it satisfies the equality (2.3). The twice continuously differential function is called the **classic state** of this system, if it satisfies the boundary problem (2.1), (2.2).*

Hence, the generalized solution of the problem or rather the generalized state of the system has a direct physical sense. It is not only superior to the classical solution, but even surpasses it from the physical point of view. This is very good. However, we do not know if it is constructive as yet. We would like to have the method of solving the problem (2.3) without its transformation to the boundary problem (2.1), (2.2). If we find it, we can decide that the generalized model is, in reality, independent from the classic one.

The formula (2.3) contains the operation of the differentiation and the integration. The integration of the integrable function is the natural operation. However, an approximation of the derivatives of the non-smooth integrable functions is not clear. Hence we shall be precise about the sense of the derivatives at the formula (2.3).

2.3   Generalized derivatives

We know that the standard method of solving the boundary problem (2.1), (2.2) is based on the approximation of the derivatives (see Chapter 1). The definition of the formulas of approximated differentiation applied the smoothness of the considered functions. However, the equality (2.3) involves the generalized derivatives but not classic ones because the element of the Sobolev space can be a non-differentiable function. Therefore, we would like to approximate the generalized derivatives for a practical determination of the generalized state of the system. First, remember the definition of the generalized derivative[6].

Definition 2.4 *The object du/dx is called the **generalized derivative** of the function u on the interval (a, b), if it satisfies the equality*

$$\int\limits_a^b \frac{du(x)}{dx}\lambda(x)dx = -\int\limits_a^b u(x)\frac{d\lambda(x)}{dx}dx \qquad (2.7)$$

for all smooth enough functions λ with zero values at the points a and b.

Remark 2.9 We shall determine the class of objects, where the generalized derivative makes sense, at Chapter 8. The definition of the generalized derivative will be clarified there. We shall also determine exactly the functional class of the function λ.

Remark 2.10 The classic derivative of the function is determined at first locally at the concrete point. Then, this can be definite as a function after variation of the point of differentiation. The generalized derivative is determined globally as the whole object.

Consider examples.

Example 2.1 *Smooth function.* Determine the function

$$u(x) = \frac{1}{2}x|x|.$$

on an interval (a, b), where $a < 0 < b$. Using equality (2.7), we get

$$\int_a^b \frac{du(x)}{dx}\lambda(x)dx = -\frac{1}{2}\int_a^b x|x|\frac{d\lambda(x)}{dx}dx =$$

$$\frac{1}{2}\int_a^0 x^2\frac{d\lambda(x)}{dx}dx - \frac{1}{2}\int_0^b x^2\frac{d\lambda(x)}{dx}dx =$$

$$-\int_a^0 x\lambda(x)dx + \int_0^b x\lambda(x)dx = \int_a^b v(x)\lambda(x)dx$$

Thus, the generalized derivative of the differentiable function u is the continuous function

$$v(x) = |x|.$$

Note, that this generalized derivative is equal to the classic one for this case. □

Example 2.2 *Non-smooth continuous function.* Find the generalized derivative of the non-smooth function v. We have

$$\int_a^b \frac{dv(x)}{dx}\lambda(x)dx = -\int_a^b |x|\frac{d\lambda(x)}{dx}dx = \int_a^0 x\lambda(x)dx - \int_0^b x\lambda(x)dx =$$

$$-\int_a^0 \lambda(x)dx + \int_0^b \lambda(x)dx = \int_a^b w(x)\lambda(x)dx,$$

where the function w is equal to 1 for the negative values of argument, and to 1 for its positive values. Therefore, the generalized derivative of the non-differential function v is the discontinuous function w. □

Note that the generalized derivative characterizes the velocity of the function change as the classic one (see Figure 2.6). For all negative values of the argument, the function v decreases with the constant velocity. Its generalized derivative w is equal to the negative constant. For all positive values of the argument, the function v increases with the constant velocity. Its generalized derivative w is equal to the positive constant. Besides, the point of the non-smoothness of the function v is the point of the discontinuity of the function w.

In reality, the formula (2.7) is applicable for the discontinuous functions too.

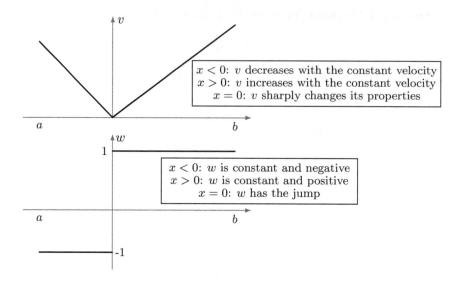

FIGURE 2.6: The generalized derivative w characterizes the velocity of the change of the function v.

Example 2.3 *Discontinuous function.* Try to find the generalized derivative of the function w. We get

$$\int_a^b \frac{dw(x)}{dx}\lambda(x)dx = \int_a^0 \frac{d\lambda(x)}{dx}dx - \int_0^b \frac{d\lambda(x)}{dx}dx = 2\lambda(0) \ \forall\lambda.$$

Hence, the generalized derivative of the function w is such an object that the integral of its product by the arbitrary smooth function is its doubled value at zero. After division by 2 this object is called the δ-*function*[7]. This is determined by the following equality with arbitrary continuous function λ

$$\int_a^b \lambda(x)\delta(x-x_0)dx = \begin{cases} \lambda(x_0), & \text{if } x_0 \in (a,b), \\ 0, & \text{if } x_0 \notin (a,b). \end{cases}$$

☐

Thus, the generalized derivative of the function w is the object y determined by the formula

$$y(x) = 2\delta(x).$$

This is not a function even. Indeed, the generalized derivative described the velocity of the change of the function. Our function w is equal to -1 for all negative arguments and +1 for all positive arguments. Therefore, its velocity

TABLE 2.3: Properties of the function and its generalized derivatives

object	definition	smoothness	monotony
u	$x\lvert x\rvert/2$	differentiability	increase
u'	$\lvert x\rvert$	non-differentiability at zero	decrease, if $x < 0$ increase, if $x > 0$
u''	-1, if $x < 0$ 1, if $x > 0$	jump at zero	piecewise constant
u'''	$2\delta(0)$	distribution	zero for all points except zero

of the change y is zero for all non-zero arguments. However, for any small value ε the function w has a finite increase 2 on the small interval $[\varepsilon, \varepsilon]$. Hence, the velocity of its change y is infinite. Of course, there are not any functions with these properties. This is an element of the class of the **generalized functions** or **distributions**[8] (see Chapter 8).

Note that the generalized derivative of the discontinuous function exists too. Of course, this is not a function. However, we can approximate it by smooth functions (see Figure 2.7). It is important that the standard sense of the classical derivative is saved in this case too. Indeed, the function w has the jump at zero. Therefore, its velocity of the change is unbounded on the neighbourhood of this point.

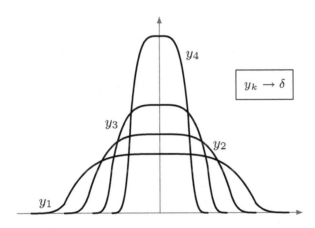

FIGURE 2.7: δ-function is a limit of a sequence of the regular functions.

Remark 2.11 We shall talk about the approximation of the δ-function at Chapter 8.

The properties of the function u and its generalized derivatives are given in Table 2.3.

In principle, we can use Definition 2.4 for the determination of the generalized derivative of the distribution.

Example 2.4 *Distribution*. Find the generalized derivative of the object y. Using the definition of the δ-function, we have

$$\int_a^b \frac{dy(x)}{dx}\lambda(x)dx = -\frac{1}{2}\int_a^b 2\delta(0)\frac{d\lambda(x)}{dx}dx = 2\frac{d\lambda(0)}{dx} \quad \forall\lambda.$$

This equality determines a generalized derivative of y.□

Note, that we do not know the direct definition of y and its generalized derivative z. However, we can interpret y as the transformation of the arbitrary smooth enough function λ to the number $2\lambda(0)$. This is can be denoted by the equality

$$\langle y, \lambda \rangle = 2\lambda(0) \; \forall\lambda.$$

This is typical for the generalized method, in principle. We cannot determine the state of the system, but we can find the response of the system to each exterior signal.

Analogically, we can interpret the object z as the transformation of the arbitrary smooth enough function λ to the number $-2d\lambda(0)/dx$. This is can be denoted by the equality

$$\langle z, \lambda \rangle = -2\frac{\lambda(0)}{dx} \; \forall\lambda.$$

The transformation that maps a function to a number is called the ***functional***. Therefore, y and z are the functionals.

Determine the properties of these functionals. For all smooth enough functions λ and μ, for all numbers a and b the map y transforms the function $a\lambda + b\mu$ to the number $2a\lambda(0) + 2b\mu(0)$. We have the equality

$$\langle y, a\lambda + b\mu \rangle = a\langle y, \lambda \rangle + b\langle y, \mu \rangle.$$

Thus, the functional y is linear. Now suppose the convergence $\lambda_k \to \lambda$ in the space of the continuous functions. Then $\lambda_k(0) \to \lambda(0)$. Therefore, we have the convergence $\langle y, \lambda_k \rangle \to \langle y, \lambda \rangle$; i.e. the functional y is continuous. Thus, the object y is the ***linear continuous functional*** on the class of the continuous functions on the given interval.

Remark 2.12 The set of all linear continuous functionals on a space is called the ***adjoint space*** (see Chapter 3). We shall determine that each distribution is a linear continuous functional on a class of smooth enough functions (see Chapter 8).

We can determine analogically, that the object z is the linear continuous functional on the class of the continuously differentiable functions.

Remark 2.13 We shall determine that each distribution is a linear continuous functional on a class of smooth enough functions (see Chapter 8).

Thus, we know, what this generalized derivative is. Now we can consider the approximation method for the analysis of the generalized model.

2.4 Approximation of the generalized model

We approximated the state equation (2.1) using the standard formulas of the approximate differentiation (see Chapter 1). However, it used the smoothness of the functions. We would like to obtain the practical algorithm of finding the generalized state of the system. Therefore, one thing is necessary to approximate the generalized derivatives.

Consider the equality (2.7)

$$\int_a^b \frac{du(x)}{dx}\lambda(x)dx = -\int_a^b u(x)\frac{d\lambda(x)}{dx}dx,$$

where λ is a smooth enough arbitrary function. Divide the interval (a, b) by equal parts with the step $h = (b - a)/M$. Approximate the integrals of the previous equality by the *right rectangle formula*[9] (see Figure 2.8)

$$\int_a^b y(x)dx \approx \sum_{i=1}^M y(x_i)h.$$

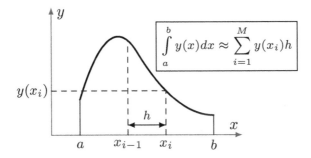

FIGURE 2.8: Right rectangle formula.

The derivative of the function λ can be approximated, for example, by the back-difference formula (see Chapter 1). This is substantiated because the values under the integrals are integrable, and the function λ is continuously differentiable. Using the boundary conditions, we have

$$\sum_{i=1}^M \frac{du(x_i)}{dx}\lambda_i h \approx -\sum_{i=1}^M u_i(\lambda_i - \lambda_{i-1}) =$$

$$-\sum_{i=1}^{M} u_i \lambda_i + \sum_{i=0}^{M-1} u_{i+1} \lambda_i = \sum_{i=1}^{M} \delta_x u_i \lambda_i h,$$

where $u_i = u(x_i)$, $\lambda_i = \lambda(x_i)$, $x_i = a + ih$. Therefore, we get

$$\sum_{i=1}^{M} \left[\frac{du(x_i)}{dx} - \delta_x u_i\right] \lambda_i \approx 0.$$

The function λ is arbitrary here. Choose it equal to zero for all considered points except x_i with fixed number i. Then we get

$$\frac{du(x_i)}{dx} \approx \delta_x u_i, \ i = 1, ..., M.$$

This is the well-known formula of the forward difference (see Chapter 1).

If we use the left rectangle formula for the approximation of the integral and the forward difference formula for the approximation of the derivative of the function λ, we will determine the back difference formula for the generalized derivative of the function u. Therefore, the standard formulas of the approximate differentiation are applicable for the generalized derivatives too. Now we can approximate the generalized model (2.3).

Approximate the integrals and derivatives of the equality

$$-\int_0^L k(x)\frac{d\lambda(x)}{dx}\frac{du(x)}{dx}dx = \int_0^L \lambda(x)f(x)dx.$$

We obtain

$$-\sum_{i=1}^{M} k_i \frac{\lambda_i - \lambda_{i-1}}{h}\frac{u_i - u_{i-1}}{h}h = \sum_{i=1}^{M}\lambda_i f_i h.$$

We have the equalities

$$-\sum_{i=1}^{M} k_i \frac{u_i - u_{i-1}}{h}\lambda_{i-1} = \sum_{i=1}^{M}\left(k_i \delta_{\bar{x}} u_i\right)\lambda_{i-1} =$$

$$\sum_{i=0}^{M-1}\left(k_{i+1}\delta_{\bar{x}}u_{i+1}\right)\lambda_i = \sum_{i=1}^{M}\left(k_{i+1}\delta_{\bar{x}}u_{i+1}\right)\lambda_i$$

because of the boundary conditions. Then we get

$$\sum_{i=1}^{M}\frac{k_{i+1}\delta_{\bar{x}}u_{i+1} - k_i\delta_{\bar{x}}u_i}{h}\lambda_i = \sum_{i=1}^{M}\lambda_i f_i h.$$

Choose λ_i equal to zero for all values of the indexes except one; we have

$$\delta_x\left(k_i \delta_{\bar{x}} u_i\right) = f_i, \ i = 1, ..., M - 1.$$

This is the difference equation. Particularly, this is equal to the standard equality (1.5).

We have the same results for finding the solution of the classic model and the generalized one. Thus, the generalized model can be solved by the standard finite difference method. The general state can be found directly without using the classic model. Note that the substantiation of the numerical method is easier here, because it is necessary to prove the convergence in the Sobolev space only. However, the convergence on the algorithm for the classic case requires the convergence in the space of the twice-differentiable functions. Hence, the numerical analysis of the generalized model is easier than of the classic one.

Thus, the generalized model seems preferable to the classic one. However, we return now to the problem of the substantiation of the determination of the considered model.

2.5 Validity of the generalized method

We have already made sure that all three stages of the research, i.e. the determination of the model, its qualitative and quantitative analysis, can be carried out on the basis of the generalized approach without any reference to the classic approach. Besides, the implementation of each of these stages is actually more effective for the generalized model than for the classical one (see Figure 2.9). Indeed, the justification for the passage to the limit for the definition of the mathematical model is realized under weaker restrictions on the state function and the parameters of the system. Therefore, the generalized model is applicable for the larger class of the systems than the classic one. The existence of the generalized state and the convergence of the numerical methods of its finding can be proved easier and for a larger class of parameters. These facts explain the extremely high popularity of the generalized approach in mathematical physics. If there exists the necessity to determine the classic solution of the system, then, as a rule, one determines its generalized solution and tries to prove its smoothness.

Remark 2.14 We shall obtain this result for the considered system in Chapter 9.

By high efficiency of the generalized approach in mathematical physics, we can suppose the possibility of a correct determination of mathematical models on its basis. Return to the construction of the generalized model from the balance relations. We have already known that the integral equality is the direct corollary of the physical laws. This does not require the use of an insufficiently validated classical model. However, the determination of the generalized model uses passage to the limit too. This operation is correct under restrictions to all considered functions. We have the possibility to choose the

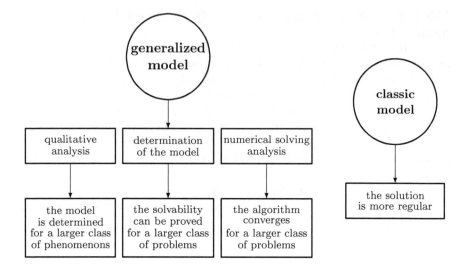

FIGURE 2.9: Advantage of the generalized and classic models.

functional properties of the given functions k and f and the arbitrary function λ. However, we do not have any information about the generalized state u. Moreover, we cannot guarantee its existence.

Of course, now we have the weaker restriction for the state function. Indeed, the generalized state should be an element of the Sobolev space only, and the classic one should be a twice continuously differentiable function. Unfortunately, all functional properties of the state function can be obtained on the second step of the analysis, i.e. after the determination of the mathematical model. Therefore, we have the same difficulty as before (see the classical case, Chapter 1).

Thus, overcoming difficulties in determining mathematical models requires the development of a qualitatively different approach to solving this problem. We would like to prove the convergence at the balance relations as the elementary volume shrinks to a point. Therefore, it will be necessary for us to consider seriously the passage to the limit that is one of the most important and difficult mathematical operations.

2.6　Conclusions

1. A generalized approach in mathematical physics is an alternative to the classical approach.

2. The generalized solution of the mathematical physics problem is an element of the Sobolev space that satisfies an integral equality.

3. The classic solution of the mathematical physics problem is always its generalized solution.

4. The smooth enough generalized solution of the mathematical physics problem is its classic solution.

5. The integral equality that determines the generalized solution is the direct corollary of physical law; therefore, it can be interpreted as a special form of mathematical model.

6. The generalized state of the system can be found directly by standard numerical methods without using the classic mathematical model.

7. The generalized method of the determination of the mathematical model is not substantiated because of the difficulties with passage to the limit.

Thus, it is necessary to find another method of analysis. We have serious problems with substantiation of the convergence. Therefore, our next step will be the analysis of the general methods of the passage to the limit.

Notes

[1]Generalized solutions for the partial differential equations are considered, for example, in [101], [102], [103], [114], [115].

[2]The fundamental lemma of the calculus of variations is the classic result of the calculus of variations; see, for example, [50], [58], [78].

[3]The mean value theorem for integrals is described, for example, in [40], [182], [189].

[4]The **black box** is a device, system or object which can be viewed in terms of its inputs and outputs (or transfer characteristics), without any knowledge of its internal workings, see [201].

[5]The problems of quantum mechanics are considered, for example, in [35], [107], [119].

[6]The generalized derivatives were be determined by *Sergey Sobolev*; see, for example, [95], [147], [158], [196].

[7]The delta function was introduced by the physicist *Paul Dirac*. This is used to model the density of a point mass or point charge as a function equal to zero everywhere except for zero and whose integral over the entire real line is equal to one.

[8]The **generalized functions** or **distributions** were determined by of *Sergey Sobolev* in 1935 in connection with the work on second-order hyperbolic partial differential equations and shock wave. The completely theory of distribution was be developed by *Laurent Schwartz* in 1950.

[9]The right rectangle formula and other formulas of numerical integration are considered, for example, in [15], [79], [151], [178].

Part II

Sequential method

We have serious difficulties with substantiation of the convergence in balance relations that are the basis of the determination of mathematical models. Therefore, overcoming the difficulties that arise requires the use of the theory of limits. It would be useful to obtain as much information on the procedure for passing to the limit, in principle. We consider, at first, the different forms of the convergence. It is very important that all these definitions are not constructive. Then serious difficulties arise in their practical application. However, we can use the Cauchy criterion as the general practical method of proving the convergence of the sequences. Unfortunately, the applicability of the Cauchy criterion is limited for the complete spaces. However, we can use the completion technique for the non-complete spaces. This gives the method of the analysis the convergence for the general case.

Chapter 3

Convergence and Cauchy principle

The cause of the difficulties of mathematical models determination is the passage to the limit. Particularly, a priori properties of the state function are required for passing to the limit in the balance relations characterizing the corresponding physical law. We consider the general definitions of the convergence for overcoming this obstacle.

One knows that a sequence tends to its limit, if all its elements with large enough numbers are close enough to this limit. Thus, an a priori knowledge of the limit is required here for establishing the fact of convergence. This situation strongly resembles the previously described difficulty, because, in applications we do not know, as a rule, the limit in advance. We can assume that the constructive methods of substantiating the passage to the limit can help us obtain the correct construction of mathematical models.

The practical method of proving the convergence is based on the Cauchy criterion. This method uses the notion of the fundamental sequence. It is most important here that the definition of the fundamental sequence includes the elements of the given sequence only, not the limit. We consider the Picard method for differential equations as an application of the Cauchy criterion. The well-known Banach fixed-point theorem is an extension of this result.

3.1 Definitions of the convergence

The limit is the most important notion of mathematical analysis[1]. Let us have a sequence $\{u_k\}$. We would like to know the properties of the element u_k after the unbounded increment of the natural number k. Maybe u_k is an approximate solution of a problem; and we would like to find its exact solution as the limit of the given sequence. Maybe we prove the solvability of a problem; and we try obtaining this result by finding the limit with special properties.

Remember the classic definition of the limit.

Definition 3.1 *The numerical sequence $\{u_k\}$ **tends** to a number u that is called its **limit**, if for any $\varepsilon > 0$ there exists a natural number $k = k(\varepsilon)$ such that $|u_k - u| < \varepsilon$ for any k that is greater than $k(\varepsilon)$.*

Unfortunately, this definition is not applicable for our situation, because we consider the convergence of functions, not of numbers. Try to extend this definition to the sequences of functions. At first, it is known that the sequence of continuous functions $\{u_k\}$ on the interval $[a, b]$ tends to a function u, if for any $\varepsilon > 0$ there exists a natural number $k(\varepsilon)$ such that

$$\max_{t \in [a,b]} \left| u_k(t) - u(t) \right| < \varepsilon$$

for any $k > k(\varepsilon)$. This inequality can be changed by another condition

$$\int_a^b \left| u_k(t) - u(t) \right| dt < \varepsilon.$$

There are the partial cases of the convergence in the linear normed spaces.

Remark 3.1 We shall consider the set of continuous functions with these forms on convergence in Chapter 4.

The **linear normed space**[2] or **normed vector space** is the set with standard linear operations of the addition and the multiplication by the arbitrary number, where for any element u it is possible to determine its norm $\|u\|$ that is a non-negative number.

Remark 3.2 Of course, this definition is non-strict, because we did not give the definitions of the linear operations and the norm. However, this is sufficient for this stage of our next analysis. We shall give the strict definition of the linear normed space in Chapter 5 (see Definition 5.5).

Remark 3.3 Sometimes we shall use the denotation $\|u\|_X$. This is the norm of the element u with respect to the concrete space X.

Remark 3.4 We consider the linear normed spaces over the real numbers only; i.e. one admits the multiplication by the real numbers. One considers also the linear normed spaces over the complex numbers with the multiplication by the complex numbers. However, this operation is not important for us.

Determine examples of norms that are important for our analysis. The set of real numbers $\mathbb{R}$ is the linear normed space with norm

$$\|u\| = |u|.$$

We can determine the norm in the Euclid space $\mathbb{R}^n$ by the formula

$$\|u\|^2 = \sum_{i=1}^{n} |u_i|^2,$$

where u_i is the i-th component of the vector u. The norms of the set of the continuous functions on the interval $[a, b]$ can be determined by the equalities

$$\|u\| = \max_{t \in [a,b]} |u(t)|,$$

$$\|u\| = \int_a^b |u(t)| dt.$$

We consider also the space $L_2(a, b)$ of the square integrable functions in the sense of Lebesgue with the norm determined by the equality

$$\|u\|^2 = \int_a^b |u(t)|^2 dt.$$

Determine also the norm on the Sobolev space $H_0^1(a, b)$ by the equality

$$\|u\|^2 = \int_a^b \left|\frac{du(t)}{dx}\right|^2 dt.$$

It is very important that the norm of the difference of two functions characterizes the closeness of these functions (see Figure 3.1). Then we determine the convergence in the arbitrary linear normed space.

Remark 3.5 We shall consider also other linear normed spaces, see Chapter 8.

Definition 3.2 *A sequence $\{u_k\}$ of a linear normed space **tends** to a **limit** u, if for any $\varepsilon > 0$ there exists a number $k = k(\varepsilon)$ such that $\|u_k - u\| < \varepsilon$ for any k that is greater than $k(\varepsilon)$.*

Of course, Definition 3.1 is the partial case of Definition 3.2, because this is the convergence of the linear normed spaces of real numbers. Both forms of convergence for the continuous functions are the convergence with respect to the space $C[a, b]$ of continuous functions with the considered norms.

Note that the convergence is based on the closeness of the elements u_k of the sequence to the limit u. This property does not have any direct connection with algebraic operations. Therefore, it will be preferable to sometimes use the convergence for sets without any operations. Particularly, we can determine the convergence on the **metric space**[3]. This is the set, where one has the possibility to determine a distance $\rho(u, v)$ between arbitrary elements u and v by the functional ρ that is called the **metric**.

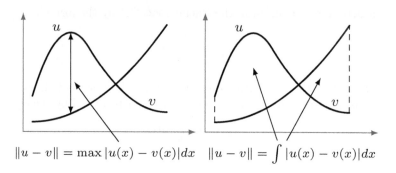

$$\|u - v\| = \max |u(x) - v(x)| dx \qquad \|u - v\| = \int |u(x) - v(x)| dx$$

FIGURE 3.1: Degree of the closeness of continuous functions.

Remark 3.6 We shall give the exact definition of the metric space soon (see Definition 3.11).

Remark 3.7 The metric is the functional on the set of element pairs (u, v) of the given set. The definition of the pair will be given in Chapter 4.

Definition 3.3 *A sequence* $\{u_k\}$ *of a metric space* ***tends*** *to an element* u, *if for any* $\varepsilon > 0$ *there exists a number* $k = k(\varepsilon)$ *such that* $\rho(u_k, u) < \varepsilon$ *for any* k *that is greater than* $k(\varepsilon)$.

The metric convergence is the convergence to zero of the numerical sequence $\{\rho(u_k, u)\}$. Therefore, the distance between the element u_k of the sequence and its limit tends to zero here (see Figure 3.2). We can determine the metric

$$\rho(u, v) = \|u - v\|$$

for any linear normed space. Thus, the convergence for the linear normed spaces is the partial case of the metric convergence.

Remark 3.8 We shall consider the strange enough metric on the set of rational numbers (see Chapter 6). It will be the basis for determining the p-adic numbers.

The metric convergence is the general enough notion. However, sometimes the closeness of elements is not determined by a metric. Consider a linear normed space X with ***scalar product***[4]. This is the ***unitary space***. For all elements u and v here there exists its scalar product (u, v) that is a number with concrete properties. Particularly, the scalar square (u, u) is non-negative. Besides, it is equal to zero for the zero element only.

Remark 3.9 We shall give the exact definition of the unitary space in Chapter 5 (see Definition 5.6).

Remark 3.10 The scalar product is the functional on the pairs set of elements of the considered set X as the metric. The definition of the pair will be given in Chapter 4.

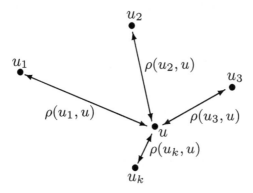

FIGURE 3.2: Elements of the sequence approach to the limit.

Consider the most important examples for us. The scalar product in $\mathbb{R}$ is the usual product. The scalar product in the Euclid space $\mathbb{R}^n$ is

$$(u, v) = \sum_{i=1}^{n} u_i v_i.$$

The spaces $C[a, b]$ and $L_2(a, b)$ are unitary with the scalar product

$$(u, v) = \int_a^b u(x)v(x)dx.$$

The Sobolev space $H_0^1(a, b)$ is unitary too with the scalar product

$$(u, v) = \int_a^b \frac{du(x)}{dx} \frac{dv(x)}{dx} dx.$$

Each unitary space is the linear normalized space with the norm

$$\|u\| = \sqrt{(u, u)}.$$

The relation between the norm and the scalar product of the unitary space X is described also by the **Schwarz inequality**[5]

$$(u, v) \leq \|u\|\|v\| \quad \forall u, v \in X.$$

Using the norm, we could consider the convergence of the sequences of the unitary spaces that is determined by its norm. This is called the **strong convergence**. However, there exists another form of convergence here[6].

Definition 3.4 *A sequence $\{u_k\}$ of a unitary space X **converges weakly** to an element u, if*

$$(\lambda, u_k) \to (\lambda, u) \quad \forall \lambda \in X.$$

Particularly, $u_k \to u$ weakly in $L_2(a, b)$, if

$$\int_a^b \lambda(x) u_k(x) dx \to \int_a^b \lambda(x) u(x) dx.$$

Analogically, $u_k \to u$ weakly in $H_0^1(a, b)$, if

$$\int_a^b \frac{d\lambda(x)}{dx} \frac{du_k(x)}{dx} dx \to \int_a^b \frac{d\lambda(x)}{dx} \frac{du(x)}{dx} dx.$$

Suppose we have the convergence $u_k \to u$ strongly in a unitary space. Determine the value

$$\left| (\lambda, u_k) - (\lambda, u) \right| = \left| (\lambda, u_k - u) \right| \le \|\lambda\| \|u_k - u\|,$$

by the Schwarz inequality. Therefore, the weak convergence follows from the strong one with the same limit. However, the inverse assertion is not obvious.

Remark 3.11 The first equality of the last formula is the corollary of the linearity of the scalar product (see Definition 5.6). Particularly, the difference between the scalar products of an element λ by elements v and u is equal to the difference of scalar products (λ, v) and (λ, u).

Example 3.1 *Weak convergence in $L_2(a, b)$.* Consider the functions

$$u_k = \frac{2}{\pi} \sin kx, \quad k = 1, 2, \dots$$

of the space $L_2(0, \pi)$. Find the value

$$(\lambda, u_k) = \frac{2}{\pi} \int_0^\pi \lambda(x) \sin kx \, dx$$

for an arbitrary function λ of $L_2(0, \pi)$. This is the *Fourier coefficient* λ_k of the function λ. By the convergence of the *Fourier series*[7]

$$\lambda(x) = \sum_{k=1}^\infty \lambda_k \sin kx,$$

it follows that $\lambda_k \to 0$. Then we have the convergence $(\lambda, u_k) \to 0$ for all λ.

Denote by u the function of the considered space that is equal to zero. Of course,

$$(\lambda, u) = \int_0^\pi \lambda(x)u(x)dx = 0.$$

Therefore, we have $(\lambda, u_k) \to (\lambda, u)$ for any λ that is the convergence $u_k \to u$ weakly in $L_2(0, \pi)$. Now calculate the norm

$$\|u_k - u\|^2 = \left(\frac{2}{\pi}\right)^2 \int_0^\pi \sin^2 kx\, dx = \frac{2}{\pi^2} \int_0^\pi (1 - \cos 2kx)\, dx = \frac{2}{\pi}.$$

Then, we do not have the convergence $u_k \to u$ strongly in $L_2(0, \pi)$. Thus, the weak convergence is weaker than the strong one. $\square$

The weak convergence in the functional spaces is not **metrisable**; i.e. it is impossible to describe by any metric[8]. This is a partial case of topological convergence.

Remark 3.12 More exact, this property is true for infinite dimensional spaces. We do not give the definition of the these spaces. However, note that the spaces $\mathbb{R}$ and $\mathbb{R}^n$ are finite dimensional; but the spaces $C[a, b]$, $L_2(a, b)$, and $H_0^1(a, b)$ are infinite dimensional. We shall consider other spaces with the weak convergence in Chapter 8. It will be the space of infinite differentiable functions and the space of distributions.

The **topological space** is the most general space, where the closeness has the sense[9]. The element of the topological space that is called the **point** has a class of **neighbourhoods**. We can describe the closeness of the points of the topological space by neighbourhoods. Particularly, the point u is **close enough** to the point v, if u belongs to a neighbourhood of v. Therefore, we can determine the convergence in the topological spaces.

Definition 3.5 *A sequence $\{u_k\}$ of a topological space **tends** to a point u, if for any neighbourhood U of u there exists a number $k(U)$ such that $u_k \in U$ for any k that is greater than $k(U)$.*

This is the **convergence with respect to the topology** of the space X.

Remark 3.13 There exists also the convergence in filter theory[10]. Note also the convergence of nets that is a non-countable analogue of the sequence. This is called the Moore–Smith convergence[11]. However, we do not consider it.

The sequence converges, if all its elements with large enough numbers belong to the arbitrary neighbourhood of the limit (see Figure 3.3). The neighbourhood of a point u of a metric space can be determined by the set of all points v that satisfy the inequality $\rho(u, v) < \varepsilon$ for an arbitrary positive number ε (see Figure 3.4). The distance between a point from this neighbourhood and the given point u is less than ε. Therefore, the metric convergence is the partial case of the topological one.

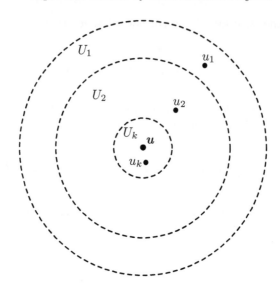

FIGURE 3.3: Elements of the sequence belong to the neighbourhoods of the limit.

The neighbourhood of the point u of a unitary space X can be determined by the set of all points v that satisfy the inequality

$$\left|(\lambda, u - v)\right| < \varepsilon \;\; \forall \lambda \in X$$

for an arbitrary positive number ε. Hence, the weak convergence of the unitary spaces is the partial case of the topological one too.

Definition 3.6 *The functional λ on the set X with linear operations is called the **linear functional**, if the following equality holds*

$$\lambda(au + bv) = a\lambda(u) + b\lambda(v) \;\; \forall u, v \in X, \forall a, b \in \mathbb{R}.$$

The map $u \to (\lambda, u)$ on a unitary space determines the linear functional that is also continuous[12]. Particularly, from the strong convergence $u_k \to u$ it follows that $(\lambda, u_k) \to (\lambda, u)$. Thus, the neighbourhoods of the unitary spaces can be determined by the linear continuous functionals. One can describe the neighbourhoods for the linear normalized spaces by the linear continuous functionals too. Moreover, we can determine the similar neighbourhoods for its extension that is the **linear topological space** or **topological vector spaces**. These are the topological spaces with continuous linear operations[13].

Remark 3.14 We will consider this class of spaces in Chapter 8.

Of course, each linear normalized space is linear topological. Denote by

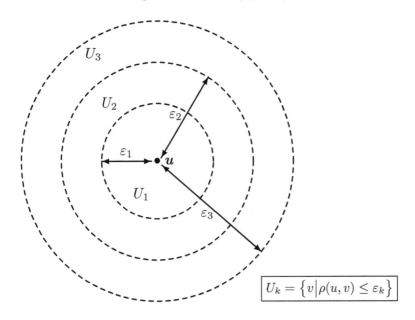

$$U_k = \left\{ v \,\middle|\, \rho(u,v) \le \varepsilon_k \right\}$$

FIGURE 3.4: Neighbourhoods of the point u of a metric space.

$\langle \lambda, u \rangle$ the value of the linear continuous functional λ at the point u of the linear topological space X. The neighbourhood of a point u here is the set of all points v that satisfy the inequality

$$\left| \langle \lambda, u - v \rangle \right| < \varepsilon \quad \forall \lambda \in X'$$

for an arbitrary positive number ε, where X' is the set of all linear continuous functionals on the space X that is called the **adjoint space** to the space X. Then we can determine the weak convergence in the linear topological spaces, particularly in the linear normalized spaces.

Definition 3.7 *A sequence $\{u_k\}$ of a linear topological space **tends** to a point u, if $\langle \lambda, u_k \rangle \to \langle \lambda, u \rangle$ for all linear continuous functional λ.*

Of course, the weak convergence of the linear topological spaces is topological convergence too. Note that the weak convergence in the functional (more exact, infinite dimensional) spaces is not metrisable; i.e. we cannot any possibly describe it by any metric, even more so, by a norm. Convergence in the sense of the norm of the linear normalized space is called **strong convergence**. If we have strong convergence in the linear normalized space, we have weak convergence to the same limit here. However, it is possible to have weak convergence without strong convergence.

TABLE 3.1:   Areas of the classic and generalized approaches

area	classic approach	generalized approach
mathematical model	classic model	generalized model
solution of the equation	classic solution	generalized solution
state of the system	classic state	generalized state
derivative of the function	classic derivative	generalized derivative
convergence of the sequence	strong convergence	weak convergence

TABLE 3.2:   Characteristics of the classic and generalized approaches

characteristic	classic method	generalized method
definition	direct	non-direct presence of arbitrary parameters
relation	each classic notion is generalized notion	each regular enough generalized notion is classic notion
property of object	stronger	weaker
class of applicability	smaller	larger

Remark 3.15 We shall consider the space of infinite differentiable functions (see Chapter 8), where we can determine the weak convergence without strong convergence.

Note the analogy between the weak convergence and the generalized approach of the analysis. Indeed, for both situations we have the relations with arbitrary values of parameters. However, the strong convergence is similar to the classic approach because for both situations we have the direct relations without arbitrary values of parameters (see Table 3.1 and Table 3.2).

The hierarchy of different classes of convergence is rendered in Figure 3.5.

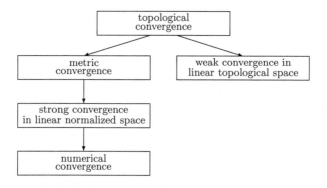

FIGURE 3.5: Hierarchy of the convergence.

We considered the different forms of the convergence. Now we try to determine how we can prove the convergence of the sequence for the practical situation.

3.2 Non-constructiveness of the limit

All considered definitions affirm the convergence of the given sequence $\{u_k\}$ to the concrete element u. We can apply it if we know the element u that can be the limit of this sequence by our supposition. Unfortunately, we do not know it, as a rule, for the practical situation. Moreover, we do not often know even the fact of the convergence. We have only the algorithm for solving a problem that is the means of the definition of the sequence. The value of the limit can be obtained as a solution of this problem after the passage to the limit. Besides, if we have already found the limit, we do not have any necessity to analyze the convergence of the sequence. The definitions of the convergence are not applicable for the practical proving of the fact of the convergence.

Remark 3.16 It is clear for the computer analyses of the sequence. Indeed, we have an algorithm for the determination of the sequence and the admissible error. However, we do not know the limit. Therefore, we do not have the effective method of interrupting the calculation, because we do not know when the difference between an element of sequence and the result becomes less than this error.

We need to know in advance the initial limit for proving the convergence of the sequence. We need to know in advance the properties of the state function for the determination of the mathematical model (see Figure 3.6). For both situations, we have the necessity to use the properties of the final object before the definition of this object. Therefore, we hope that these problems have similar resolutions. If we find the constructive method of proving the convergence, then we obtain perhaps the chance to find the method of correct determination of the mathematical model.

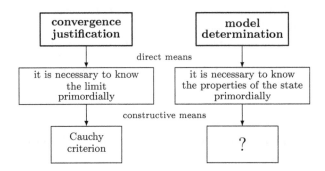

FIGURE 3.6: Determination of the limit and the mathematical model.

Now we consider the constructive method of the determination of the convergence without a priori information about the limit.

TABLE 3.3: Convergence and fundamentality of the sequence

property	convergence	fundamentality
definition	approaching of the sequence elements to a limit	approaching of the sequence elements
a priori information	knowledge of the limit	no
computer experiment	experiment estimate of the convergence is impossible	experiment estimate of the fundamentality is possible

3.3   Cauchy criterion of the convergence

The basic method of the practical proof of the convergence of the numerical sequences is based on the notion of the fundamental sequence.

Definition 3.8 *The sequence of real numbers $\{u_k\}$ is called **fundamental** or **Cauchy sequence**, if for any value $\varepsilon > 0$ there exists a number $k(\varepsilon)$, such that $|u_n - u_m| < \varepsilon$ for all m and n greater than $k(\varepsilon)$.*

The basis of our theory is the ***Cauchy criterion***[14].

Theorem 3.1 *Any fundamental sequence of real numbers converges.*

Remark 3.17 We shall prove the Cauchy criterion in Chapter 5 (see Lemma 5.9).

We do not have any necessity to know the limit of the sequence for the justification of its fundamentality. We use the elements of the sequence only here (see Table 3.3). The effectiveness of the Cauchy criterion is the possibility to prove the convergence if we have the fundamental sequence. Of course, this result does not determine the value of the limit. However, we often have an interest in the fact of the convergence only.

Consider the easiest application of the Cauchy criterion.

Example 3.2 ***Numerical sequence.*** Let us consider the sequence $\{u_k\}$ that is determined by the equality

$$u_k = 1/k, \ k = 1, 2, \dots .$$

We would like to know, if this sequence converges or not. Determine $n = m+p$, where $p > 0$. We have

$$\left| u_n - u_m \right| = \frac{p}{m(m+p)} = \frac{1}{m}\frac{1}{m/p+1} \leq \frac{1}{m}.$$

Choose the number m such that the inequality $m > 1/\varepsilon$ holds for the arbitrary fixed value ε. Then the elements u_n and u_m are close enough. Therefore, we have the fundamental sequence. Using the Cauchy criterion, we prove the convergence of the sequence without the knowledge of its limit. $\square$

TABLE 3.4:   Fundamental sequences

space	fundamentality
set of rational numbers	$\|u_n - u_m\| \to 0$
linear normalized space	$\|u_n - u_m\| \to 0$
metric space	$\rho(u_n, u_m) \to 0$
linear topological space	$\left|\langle \lambda, u_n - u_m \rangle\right| \to 0 \ \forall \lambda$

The Cauchy criterion is true for the Euclid space, the space of the continuous functions with the norm the maximum of the absolute value, and some others. It applies to the general linear normed spaces and metric spaces because we can determine the notion of the fundamental sequences there.

Definition 3.9 *The sequence of elements* $\{u_k\}$ *of a linear normed space (respectively, metric space) is called **fundamental**, if for any value* $\varepsilon > 0$ *there exists a number* $k(\varepsilon)$, *such that* $\|u_n - u_m\| < \varepsilon$ *(respectively,* $\rho(u_m, u_n) < \varepsilon$*) for all* m *and* n *greater than* $k(\varepsilon)$.

Remark 3.18 The concept of the fundamental sequence does not make sense in the general topological space, because the notion of the mutual proximity of points is not defined there. There one can only estimate the proximity of one point to another. However, the fundamental sequences make sense in the linear topological spaces, where the estimate of the proximity of the elements of the sequence can be determined by the value $\left|\langle \lambda, u_n - u_m \rangle\right|$ for all linear continuous functional λ (see Table 3.4). We shall consider this case for determining the distributions (see Chapter 8). The uniform spaces[15] are a generalization of such spaces. There it is possible to indicate whether the points under consideration are sufficiently close to each other without singling out one of them. Naturally, the metric spaces are uniform, although not every uniform space is metrisable. The hierarchy of topological spaces is shown in Figure 3.7.

Cauchy criterion is a basis of many important mathematical results. Consider its application to the differential equation theory.

3.4   Picard's method for differential equations

Consider the **Cauchy problem** for the general first order **differential equation**[16]

$$\frac{du(x)}{dx} = f(x, u(x)), \ x > 0; \ u(0) = u^0, \tag{3.1}$$

where f is a given function, u_0 is a constant. We would like to prove the solvability of this problem under some assumptions about the function f. Consider the following property.

Definition 3.10 *The function* $F = F(t)$ *is called **Lipschitz continuous**, if the following inequality holds*

$$\left|F(s) - F(t)\right| \leq L|s - t| \ \forall s, t,$$

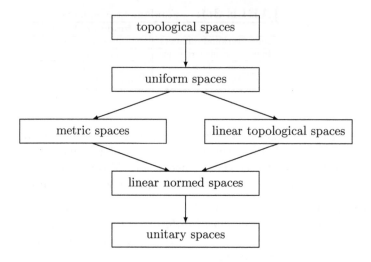

FIGURE 3.7: Hierarchy of topological spaces.

*where the positive constant L is called the **Lipschitz constant**.*

Any Lipschitz continuous function is continuous, and each differentiable function is Lipschitz continuous. However, there exist continuous functions that are not Lipschitz continuous and Lipschitz continuous functions that are not differentiable (see Figure 3.8).

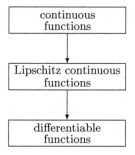

FIGURE 3.8: Hierarchy of the continuous functions.

Prove the following result that is ***Picard's theorem***[17].

Theorem 3.2 *Let the function f be continuous with respect to the first argument and Lipschitz continuous with respect to the second argument. Then there exists a positive number T such that the problem (3.1) has a unique continuous solution on the interval $(0, T)$.*

Proof. After integration of the differential equation from zero to a value x we have the equality

$$u(x) = u^0 + \int_0^x f(\xi, u(\xi))d\xi, \ x > 0. \tag{3.2}$$

Then the Cauchy problem (3.1) is equivalent to the **integral equation** (3.2). Using the **Picard method** or the **method of successive iterations**[18], we try to determine the approximate solution of this problem by the formula

$$u_{k+1}(x) = u^0 + \int_0^x f(\xi, u_k(\xi))d\xi, \ x > 0, \tag{3.3}$$

where the initial approximation $u_0 = u_0(x)$ is arbitrary.

Prove that the sequence $\{u_k\}$ is fundamental on the space $[0, T]$ of continuous functions on an interval $(0, T)$. Determine $n = m + p$; we find

$$\|u_n - u_m\| = \max_{x \in (0,L)} |u_{m+p}(x) - u_m(x)|.$$

By the equality (3.3) we get

$$|u_{m+p}(x) - u_m(x)| = \left| \int_0^x \left[f(\xi, u_{m+p-1}(\xi)) - f(\xi, u_{m-1}(\xi)) \right] d\xi \right| \le$$

$$\int_0^x \left| f(\xi, u_{m+p-1}(\xi)) - f(\xi, u_{m-1})(\xi)) \right| d\xi \le L \int_0^x |u_{m+p-1}(\xi) - u_{m-1}(\xi)| d\xi,$$

where L is the Lipschitz constant of the function f with respect to its second argument. Then we have

$$\|u_{m+p} - u_m\| = \max_{x \in (0,L)} |u_{m+p}(x) - u_m(x)| \le L \int_0^x |u_{m+p-1}(\xi) - u_{m-1}(\xi)| d\xi =$$

$$L \int_0^T |u_{m+p-1}(\xi) - u_{m-1}(\xi)| d\xi \le L \int_0^T \max_{\xi \in (0,L)} |u_{m+p-1}(\xi) - u_{m-1}(\xi)| d\xi =$$

$$LT \|u_{m+p-1} - u_{m-1}\|.$$

Repeat these transformations. We obtain

$$\|u_{m+p} - u_m\| \le (LT)^m \|u_p - u_0\|. \tag{3.4}$$

Then we find

$$\|u_p - u_0\| \le \|u_p - u_{p-1}\| + \|u_{p-1} - u_{p-2}\| + \ldots + \|u_1 - u_0\| \le$$

$$(LT + 1)\|u_{p-1} - u_{p-2}\| + \ldots + \|u_1 - u_0\| \le$$

$$\left[(LT)^{p-1} + \ldots + LT + 1\right]\|u_1 - u_0\| \le \sum_{i=0}^{\infty} (LT)^i \|u_1 - u_0\|.$$

Suppose $T < 1/L$. Using the properties of the geometric progression, we get the inequality

$$\|u_p - u_0\| \le (1 - LT)^{-1}\|u_1 - u_0\|.$$

Put this result to the inequality (3.4). We have

$$\|u_{m+p} - u_m\| \le (LT)^m (1 - LT)^{-1}\|u_1 - u_0\|.$$

The value of the right-hand side of this inequality is small enough for large enough numbers m.

Thus, the sequence $\{u_k\}$ is fundamental on the space $[0, T]$ for the small enough value T. Using the Cauchy criterion, we prove the convergence of this sequence. Therefore, there exists a continuous function u such that $u_k \to u$ in $[0, T]$ as $k \to \infty$.

Prove that the limit u is a solution of the problem (3.1). After passing to the limit at the equality (3.3) with using the continuity of the function f with respect to its second argument, we get the equality (3.2) that is equal to the Cauchy problem (3.1).

Prove now the uniqueness of this solution. Suppose on the contrary, there exist two solutions u_1 and u_2 of the considered problem. It satisfies the equality (3.3), i.e.

$$u_i(x) = u^0 + \int_0^x f(\xi, u_i(\xi))d\xi, \ i = 1, 2.$$

Find the difference

$$u_1(x) - u_2(x) = \int_0^x \left[f(\xi, u_1(\xi)) - f(\xi, u_2(\xi)) \right] d\xi.$$

Then we have

$$|u_1 - u_2| \le \left| \int_0^x \left[f(\xi, u_1(\xi)) - f(\xi, u_2(\xi)) \right] d\xi \right| \le$$

$$\int_0^x \left| f(\xi, u_1(\xi)) - f(\xi, u_2(\xi)) \right| d\xi \le L \int_0^x |u_1(\xi) - u_2(\xi)| d\xi,$$

TABLE 3.5: Proof of the Picard's theorem

step	action	result
1	algorithm implementation	$\{u_k\}$
2	proving of the fundamentality	$\|u_n - u_m\| \to 0$
3	using the Cauchy criterion	$u_k \to u$
4	analysis of the limit	u is a solution

because of the Lipschitz condition. Determine the value

$$\|u_1 - u_2\| = \max_{x \in [0,T]} \left|u_1(x) - u_2(x)\right| \le$$

$$L \int_0^x \left|u_1(\xi) - u_2(\xi)\right| d\xi \le LT\|u_1 - u_2\|.$$

By the inequality $LT < 1$, we obtain

$$\|u_1 - u_2\| < \|u_1 - u_2\|,$$

that is false. Therefore, our supposition about the existence of the non-unique solution of the Cauchy problem is false too. This completes the proof of Theorem 3.2. □

Note the steps of the proof that are typical for the practical application of the Cauchy criterion (see Table 3.5). At first, we choose the algorithm for the determination of the sequence $\{u_k\}$. Then we prove that this sequence is fundamental. By the Cauchy criterion, this sequence is convergent. After the analysis of its limit, we prove that the obtained limit is the solution of the given problem.

Consider examples.

Example 3.3 *Local solvability*[19]. We have the Cauchy problem

$$\frac{du}{dx} = u^2, \; x > 0; \; u(0) = 1.$$

Transform the given equation

$$\frac{du}{u^2} = dx.$$

After integration of this equality we obtain

$$-\frac{1}{u(x)} = x + c,$$

where c is an arbitrary constant. Then we find the general solution of the equation

$$u(x) = -\frac{1}{x + c}.$$

Using the initial condition, we find $c = -1$. Thus, the solution of the Cauchy problem is (see Figure 3.9)

$$u(x) = \frac{1}{1-x}.$$

However, this formula has the sense for $x < 1$ only. Note that the Picard's theorem guarantees the solvability of the problem on the small enough interval only that is the **local solution**. □

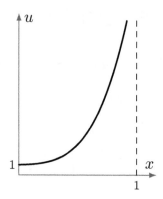

FIGURE 3.9: The Cauchy problem has a local solution only.

Example 3.4 *Non-uniqueness of the solution*[20]. Consider now the Cauchy problem

$$\frac{du}{dx} = \sqrt{|u|}, \ x > 0; \ u(0) = 0.$$

It is obvious that the function

$$u_c(x) = \begin{cases} 0, & \text{if } x \leq c, \\ (x-c)^2/4, & \text{if } x > c \end{cases}$$

is the solution of the given problem for any non-negative constant c (see Figure 3.10). The Picard theorem is not applicable for this case because the function at the right-hand side of the given equation does not satisfy the Lipschitz condition (see Figure 3.11). □

Consider an extension of Theorem 3.2.

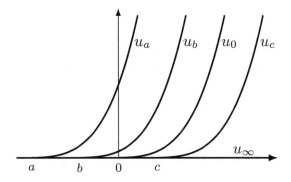

FIGURE 3.10: The Cauchy problem has many solutions.

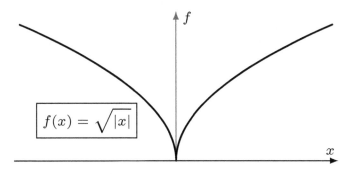

FIGURE 3.11: The function f does not satisfy the Lipschitz condition.

3.5   Banach fixed point theorem

Extend the Picard method to the general equations in the metric spaces. Therefore, we need to give the exact definitions of these spaces[21].

Definition 3.11 *Let us have a set X and a map ρ such that for all elements u and v there exists a non-negative number $\rho(u, v)$. This is the **metric space** if the following properties hold:*

i) $\rho(u, v) = 0$ *if and only if* $u = v$;

ii) $\rho(u, v) = \rho(v, u)$ $\forall u, v \in X$ *(the symmetry)*;

iii) $\rho(u, v) \leq \rho(v, w) + \rho(w, u)$ $\forall u, v, w \in X$ *(the **triangle inequality**, see Figure 3.12).*

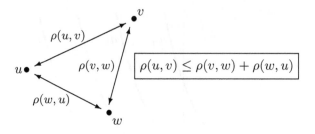

FIGURE 3.12: Triangle inequality.

Let A be an operator on a space X with a metric ρ. Consider the operator equation

$$u = Au. \tag{3.5}$$

Its partial case is the integral equation (3.3). The solution of the equation (3.5) is called the ***fixed point***[22] of the operator A. Figure 3.13 explains the geometric sense of this notion.

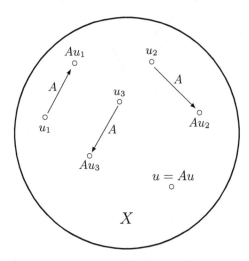

FIGURE 3.13: Fixed point of an operator A is a solution of the equation $u = Au$.

Suppose the operator satisfies the Lipschitz condition

$$\rho(u, v) \le \theta \rho(u, v) \ \ \forall u, v \in X,$$

where the positive constant θ is less than 1 (see Figure 3.14). An operator with this property is called the ***contracting operator***.

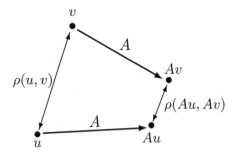

FIGURE 3.14: The operator A is contracting.

Consider the sequence $\{u_k\}$, determined by the formulas

$$u_{k+1} = Au_k. \tag{3.6}$$

We have the iterative method that is called the ***method of successive iterations***. Its geometric illustration for ***nonlinear algebraic equations*** $u = f(u)$ is given by Figure 3.15.

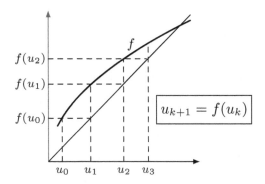

FIGURE 3.15: Iterative method for the algebraic equation.

Using the Lipschitz condition, we get

$$\rho\big(u_{m+p}, u_m\big) = \rho\big(Au_{m+p-1}, Au_{m-1}\big) \le$$

$$\theta\rho\big(u_{m+p-1}, u_{m-1}\big) \le \dots \le \theta^m \rho(u_p, u_0).$$

We have the inequality

$$\rho(u_p, u_0) \le \rho(u_p, u_{-1}) + \dots + \rho(u_1, u_0) \le (\theta^{p-1} + \dots + \theta + 1)\rho(u_1, u_0).$$

If $\theta < 1$, we obtain

$$\rho(u_{m+p}, u_m) \leq \theta^m (1 - \theta)^{-1} \rho(u_1, u_0).$$

The term at the right-hand side of this inequality is small enough for a large enough value of the number m. Therefore, our sequence is fundamental.

If the Cauchy principle is applicable, then this sequence converges. Therefore, there exists its limit u. Then after passing to the limit at the equality (3.6) we obtain the equality (3.5). Hence, the equation (3.5) is solvable; its solution is the limit of the sequence $\{u_k\}$.

We can determine also the uniqueness of the solution. Indeed, suppose there exist two solutions u_1 and u_2 of the considered equation that is

$$u_i = Au_i, \quad i = 1, 2.$$

Then we find

$$\rho(u_1, u_2) = \rho(Au_1, Au_2) \leq \theta\rho(u_1, u_2) < \rho(u_1, u_2).$$

Therefore, our supposition about the existence of the non-unique solution of the given equation is false.

We can estimate also the velocity of the convergence, at last. Suppose u is the solution of the given equation, and the sequence $\{u_k\}$ is determined by the formula (3.6). Then we have

$$\rho(u_k, u) = \rho(Au_k, Au) \leq \theta\rho(u_{k-1}, u) \leq \ldots \leq \theta^k \rho(u_0, u).$$

Thus, we have the exponential convergence of the algorithm, because $\theta < 1$.

Remark 3.19 We can have the divergence of the algorithm, if $\theta > 1$ (see Figure 3.16).

The obtained complete result is the subject of the **Banach fixed point theorem**[23]. This is the extension of the Picard theorem. This is a very important application of the Cauchy criterion.

Remark 3.20 The strong definition of this result will be given in the next chapter.

Thus, we have an effective enough method for proving convergence. Hence, we hope to use it for our general problem. However, we do not have information about the applicability of the Cauchy principle for general metric spaces. This is our next question.

3.6   Conclusions

1. The justification of the determination of the mathematical models is based on the passage to the limit.

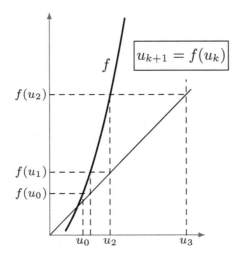

FIGURE 3.16: Iterative method for the algebraic equation.

2. The passage to the limit is substantiated there if we can guarantee positive properties of the state function.

3. The general definition of the limit is not constructive because it uses a priori knowledge of the limit.

4. There exists an analogy between the necessity to have a priori information for the determination of mathematical models and the proof of the convergence.

5. We suppose that constructive methods of proving the convergence could help to pass to the limit in the balance relations for the determination of mathematical models.

6. The general proof of the convergence can be based on the Cauchy criterion that does not use the a priori knowledge of the limit.

7. The Picard Theorem and The Banach Theorem are important applications of the Cauchy criterion.

It is necessary to check the applicability of the Cauchy criterion for the arbitrary metric space.

Notes

[1]The strict definition of the limit was be proposed by *Bernard Bolzano* and *Augustin-Louis Cauchy*. The classic theory of limits is considered in any course on mathematical analysis, see, for example, [40], [109], [182], [189].

[2]The linear normed spaces are considered by *Frigyes Riesz* and *Stefan Banach*. The theory of the linear normed spaces is the part of functional analysis, see, for example, [42], [71], [81], [91], [95], [100], [142], [147].

[3]The metric space was be determined by *Maurice Frechet*. The theory of metric spaces is described, for example, in [30], [71], [95], [135], [142].

[4]The spaces with scalar product (dot product or inner product) are considered in the standard textbooks of functional analysis, see, for example, [71], [91], [95], [100], [123], [142].

[5]The Schwarz inequality is also called the **Cauchy–Bunyakovsky inequality**. Its discrete analogue was proved by *Augustin-Louis Cauchy* in 1821. The first result for integrals was obtained by *Viktor Bunyakovsky* in 1859. The modern proof of the integral inequality was given by *Hermann Amandus Schwarz* in 1888.

[6]The weak convergence in linear normed spaces was considered by *Stefan Banach*. The properties of the weak convergence are considered in the courses of functional analysis, see, for example, [71], [81], [91], [95], [142], [147].

[7]The theory of Fourier series is the important part of mathematical analysis, which has many applications, see, for example, [71], [91], [142], [147], [188], [207].

[8]The conditions of the metrisability of the topological spaces are considered in [7], [37], [83]

[9]The topological spaces were be determined by *Felix Hausdorff*. The topological space is the fundamental notion of general topology; see, for example, [7], [26], [51], [81], [83], [98], [173].

[10]The **filter theory** is considered in [26]. This form of convergence was proposed by *Henri Paul Cartan* in 1937.

[11]The **net** theory and the **Moore–Smith convergence** are considered in [83]. This form of convergence was proposed by *Eliakim Hastings Moore* and *Herman Lyle Smith* in 1922 [48].

[12]The theory of linear continuous functionals is the direction of the functional analysis; see, for example, [71], [81], [91], [95], [142], [147].

[13]The theory of linear topological spaces is considered, for example, in [28], [93], [147], [144], [153].

[14]The fundamental sequences and Cauchy criterion were determined by *Bernard Bolzano* and *Augustin-Louis Cauchy*, see [109], [182], [189].

[15]The **uniform spaces** was determined be André Weil in 1937. These spaces are considered in [26]

[16]The theory of differential equations is described, for, example, in [13], [36], [65], [80], [138], [183], [206].

[17]The considered existence theorem for the Cauchy problem was proved by *Emile Picard*. One knows existence theorems for this problem by Cauchy, Weierstrass, Lipschitz, Peano, Bendixson, Painlevé, etc., see [13], [36], [65], [80], [138], [183], [206]. There exist numerical methods of solving the Cauchy problem for the differential equations. There are the methods of Euler, Runge and Kutta, Adams, Miln, etc., see [74], [106].

[18]One uses the method of successive iterations for nonlinear algebraic equations, the systems of linear algebraic equations, integral equations, etc.

[19]Example 3.3 is considered in [66].

[20]This example is considered in [66].

[21]The properties and the examples of metric spaces are considered, for example, in [30], [81], [83], [95], [135], [142].

[22]The fixed point theory is described in [3], [5], [87], [88], [94], [95].

[23]It is known as the Brouwer fixed point theorem, Schauder fixed-point theorem, etc., see [3], [5], [89], [90], [96], [97].

Chapter 4

Completeness and real numbers

We would like to substantiate the determination of the mathematical models. The general cause of the difficulties here is the passage to the limit in the balance relations of the elementary volume as this volume shrinks to a point. This operation requires properties of the considered functions, just as we do not obtain the equation with respect to these functions yet. The effective method of the substantiation of the convergence is necessary for this procedure. We know that the standard definitions of the limit are non-constructive, because the knowledge of the limit is necessary here. However, we often do not know the fact of the convergence.

The existence of the limit of the numerical sequences can be determined by the Cauchy criterion. This method can deduce the convergence of the sequences from its fundamentality. Note that the definition of the fundamental sequence uses the elements of this sequence only. This technique is applicable for some functional spaces too, for example, for the space of continuous functions with the standard norm. This circumstance gives us some hope, since in the process of determining the mathematical physics equations, it is necessary to work with functions, not numbers.

Unfortunately, the Cauchy criterion is not applicable for many metric spaces, particularly, for the sets of rational numbers, positive numbers, continuous functions with the integral norm, and Riemann integrable functions. We cannot guarantee the convergence of fundamental sequences there. The Cauchy criterion is applicable for complete spaces only. Moreover, the majority of the mathematical spaces are non-complete. Therefore, we need to have the effective method of passage to the limit for the non-complete spaces too.

Note that for all examples of the divergent fundamental sequences there exist points that can be interpreted as generalized limits of these sequences. These points are not the elements of the given spaces. However, they will be the points of its extensions. Thus, the non-complete spaces can be extended to complete spaces. This idea is realized for Cantor's definition of the set of real numbers.

4.1   Inapplicability of the Cauchy criterion

The Cauchy criterion is a very effective method of proving the convergence of the sequences. This is applicable not only for numerical sequences, but for classes of functional sequences too. However, it is not obviously applicable for all metric spaces. Determine the relation between the convergent and fundamental sequences.

By the triangle inequality, we have

$$\rho(x_m, x_n) \leq \rho(x_m, x) + \rho(x, x_n).$$

If the sequence $\{x_k\}$ has a limit x, then the terms at the right-hand side of this inequality tend to zero as m and n tend to infinity. Therefore, this sequence is fundamental. However, the inverse assertion is not obvious.

Consider some examples[1].

Example 4.1 *Space of rational numbers*. Consider the set of all rational numbers $\mathbb{Q}$. This is the metric space with the natural metric (and the norm too) that is the subspace of the space of real numbers. Determine the following sequence

$$x_1 = 3, \ x_2 = 3.1, \ x_3 = 3.14, \ x_4 = 3.141, \ x_5 = 3.1415, \ \dots .$$

This is the fundamental sequence because of the estimate

$$\left| x_{m+p} - x_m \right| \leq 10^{1-m} \ \forall m, p.$$

If the Cauchy criterion is applicable on the space $\mathbb{Q}$, then there exists a rational number x that is the limit of the sequence $\{x_k\}$. However, this limit does not exist, because this sequence determines the irrational number π. This is not a point of the initial space $\mathbb{Q}$. We have the sequence $\{x_k\}$ on the space of rational numbers without a rational limit. Therefore, this sequence is divergent; and the Cauchy criterion is not applicable on the metric space of rational numbers. □

Remark 4.1 It may appear that the sequence $\{x_k\}$ is convergent with the number π as the limit. However, we cannot talk about the convergence or the fundamentality of the sequence without indication of the metric space, where we consider this property. Indeed, for all sequences there exists a topological space such that this sequence is convergent there. This is, for example, the *anti-discrete topological space*[2]. If we would like to use the Cauchy criterion, then we need to consider the convergence and the fundamentality for the same space.

Remark 4.2 We shall consider the set of rational numbers with another metric, see Chapter 6.

Example 4.2 *Space of positive numbers.* Consider now the space of positive real numbers $\mathbb{R}_+$ with the same metric. This is another subspace of the space of real numbers. Determine the sequence $\{x_k\}$ by the following equality

$$x_k = \frac{1}{k}, \ k = 1, 2, \dots .$$

This is the fundamental sequence on the space $\mathbb{R}_+$ because of the equality

$$\left|x_{m+p} - x_m\right| = \frac{1}{m} - \frac{1}{m+p} = \frac{p}{m(m+p)}.$$

However, we do not have any positive number that is the limit of this sequence. Therefore, we have the divergence of the fundamental sequence on the metric space of positive numbers. □

Example 4.3 *Space of continuous functions with the integral metric.* Determine now the space $C[-1,1]$ of the continuous functions on the interval $[-1,1]$ with the integral metric

$$\rho(x,y) = \int\limits_{-1}^{1} |x(t) - y(t)|\,dt.$$

Consider the sequence of the functions $\{x_k\}$ that are determined by the equalities

$$x_k(t) = \begin{cases} t^{1/k}, & \text{if } 0 \le t \le 1, \\ -(-t)^{1/k}, & \text{if } -1 \le t \le 0, \end{cases} \quad k = 1, 2, \dots .$$

There are the continuous functions on the interval $[-1,1]$ (see Figure 4.1). It is obvious that the sequence $\{x_k\}$ is fundamental here. However, this sequence does not have any continuous function as a limit. Thus, the Cauchy criterion is not applicable for the metric space of continuous functions with the integral metric too. □

Example 4.4 *Space of Riemann integrable functions*[3]. Consider the set of Riemann integrable functions on the interval $[0,1]$. Determine here the metric by the equality

$$\rho(x,y) = \int\limits_{0}^{1} |x(t) - y(t)|\,dt.$$

The set of all rational numbers on the interval $[0,1]$ is denumerable. Therefore, it can be interpreted as a sequence $\{q_k\}$. Determine the following sequence of

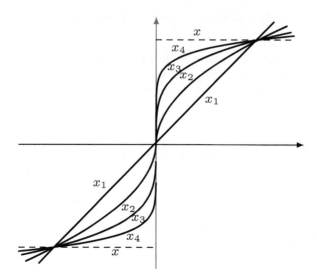

FIGURE 4.1: Divergence of the fundamental sequence of the continuous functions.

the functions $\{x_k\}$. Let x_1 be the function equal to 1 at the point q_1 with zero values for all other points. If the function x_{k-1} is known, we determine the next function x_k such that it equals 1 at the point q_k and x_{k-1} for all other points. The function x_k is integrable in the sense of Riemann (see Figure 4.2) because it is equal to zero everywhere with the exception of the finite set of points. The value $\rho(x_m, x_n)$ is equal to zero because we have the distinction between these functions on the finite set only. Therefore, the sequence $\{x_k\}$ is fundamental on the metric space of Riemann integrable functions. We could interpret the **Dirichlet function** D as the limit of this sequence. It has the value 1 for all rational arguments and the value 0 for all irrational numbers[4]

   Try to calculate its **Riemann integral**. Divide its domain by parts. Then we choose a rational point from each part. The Riemann integral sum is the sum of the areas of rectangles, where the sides are the value of D at the chosen point that is 1 and the length of the considered part of the fragmentation (see Figure 4.3). Of course, the integral sum is equal to 1. Pass here to the limit as the maximal length of these intervals tends to zero; we obtain the value 1 as the limit. However, for any part of the fragmentation we can choose an irrational point with zero value of the Dirichlet function. Therefore, the integral sum is equal to zero too (see Figure 4.3). Its limit as the maximal length of these intervals tends to zero is zero too. Thus, the limit of the integral sum depends on the choice of the point from the intervals of fragmentation. Hence, the Dirichlet function is not Riemann integrable. We determine that the fundamental sequence of the Riemann integrable functions does not have a Riemann integrable limit. Thus, the Cauchy criterion is not applicable here.□

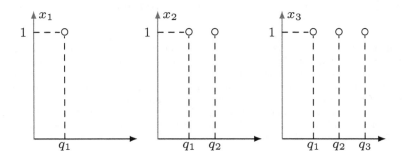

FIGURE 4.2: Fundamental sequence of Riemann integrable functions.

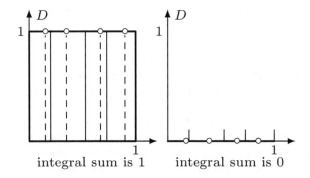

FIGURE 4.3: Integral sum depends on the choice of the points.

Remark 4.3 Other examples of incomplete spaces will be considered in the subsequent chapters.

Thus, there exist many metric spaces, where the Cauchy criterion is not applicable. These results lead us to the notion of a complete metric space.

4.2 Complete metric spaces

We have, in reality, two fundamentally different classes of metric spaces. There are the metric spaces with the Cauchy criterion and the metric spaces without the Cauchy criterion.

Definition 4.1 *The metric space is called* **complete**[5], *if each fundamental sequence converges here.*

Remark 4.4 The concept of completeness is connected with the natural desire to have unconditional fulfillment of some property, in this case, the convergence of fundamental sequences. We have already encountered a similar phenomenon when considering the problem of unique solvability of equations (see Remark 2.7). In particular, the additive equation on the set of natural numbers is "incomplete" in a certain sense, and "complete" on the set of integers. Similarly, the multiplicative equation on the set of integers can be considered "incomplete", and this is "complete" on the set of rational numbers if one does not take into account equations of the type $0 \cdot x = f$.

Thus, the Cauchy criterion is applicable for the complete spaces only. We have already known that the space of real numbers, Euclid space and the space of continuous functions with the metric determined by the maximum of the absolute value are complete. However, the spaces of rational or positive numbers, continuous functions with the integral metric and Riemann integrable functions are not complete.

Now we can give the exact determination of the ***Banach fixed point theorem***[6].

Theorem 4.1 *Let X be a complete metric space, and A be a contracting operator. Then the operator A has a unique fixed point that is the limit of the method of successive iterations for all initial approximations.*

The complete metric spaces have many applications. The most important classes of the complete metric spaces are the Banach spaces and the Hilbert spaces[7]. The ***Banach space*** is the complete linear normed space; and the ***Hilbert space*** is the complete unitary space (see Figure 4.4).

Remark 4.5 The Cauchy criterion has the sense not only for the metric spaces, but also for its extension. These are the complete uniform spaces.

Unfortunately, there exist incomplete metric spaces. Besides, each positive property is the exception as a rule. Therefore, the majority of metric spaces (and uniform spaces too) are incomplete. Hence, we would like to have an effective method of passage to the limit for the general case, namely without the property of completeness.

4.3   Completion problem

We return to consideration of the fundamental sequences without limits (see our previous examples). The divergent considered sequence of rational numbers does not have a rational limit. However, there exists an element π that is not a rational number. We can interpret it as a special weaker form

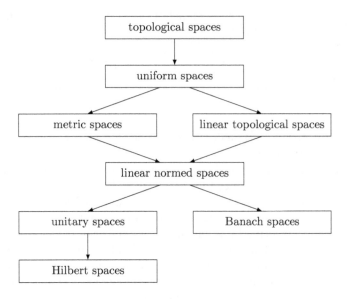

FIGURE 4.4: Hierarchy of topological spaces.

of the limit for this sequence. The divergent fundamental sequence of positive numbers has a special element 0 that is not a positive number. However, we can interpret it as a special form of the limit for the sequence of positive numbers. The analogical property is true for the sequence of continuous functions from Figure 4.1. There exists the discontinuous function

$$x(t) = \begin{cases} 0, & \text{if} \quad 0 \le t \le 1, \\ -1, & \text{if} \quad -1 \le t \le 0 \end{cases}$$

that can be interpreted as a limit of the sequence of continuous functions. Besides, the Dirichlet function can be interpreted as a special form of the limit for the considered sequence of Riemann integrable functions.

The above concepts are not strict. However, we could assume that after the adjunction of the original incomplete metric space by artificially defined "generalized limits" of divergent fundamental sequences, we provide in some sense the validity of the Cauchy convergence criterion. Thus, there is a hope that, as a result, we will get an effective way to substantiate the passage to the limit even in the absence of the completeness of the space. Of course, there may be natural fears of the legitimacy of such reasoning. Indeed, we do not have the "true" limit; and we ourselves are conceiving elements that do not really exist, which we then try to artificially interpret as the limits of the divergent fundamental sequences. However, the numbers π and zero mentioned above, the function x in Figure 4.1 and the Dirichlet function are concrete

TABLE 4.1: Analogy between definition of integer and real numbers

given set	analyzed object	difficulty	desirable result	extended set
natural numbers	additive equation	insolvability of the equation	existence of the solution	integer numbers
rational numbers	fundamental sequence	divergence of the sequence	convergence of the sequence	real numbers

mathematical objects although there are not elements of the corresponding incomplete metric spaces.

Note that all mathematical concepts were actually introduced artificially at the time when a serious need for them arose. For example, let's remember how negative numbers were defined. I had some apples. I got three more apples that resulted in five apples. It is required to find out how many apples I had originally. Denoting by x the unknown number of apples, we obtain the equation $x+3 = 5$. Solving this equation, we find the value $x = 2$. Now change the situation. I have apples. I got five more apples extra that resulted in three apples. We have the equation $x+5 = 3$. According to natural sense, the number of apples must necessarily be positive or, in extreme cases, zero. Therefore, we conclude that the equation under consideration has no solution. However, we supplement the set of natural (more precisely, non-negative integers) numbers by solutions of equation $x + a = b$ that is called the **additive equation** for all natural parameters a and b. Then our equation has the solution that can be interpreted as my duty to somebody. Of course, this does not belong to the given set of non-negative integer numbers. However, this is an element of an artificially constructed extension of this set. There is an obvious analogy between the method of determining real, in particular, irrational numbers and integers, in particular, negative numbers (see Table 4.1).

In both cases we have some procedure (the convergence of fundamental sequences, the resolution of the additive equation) that may not be realizable on a given set (rational numbers, natural numbers). However, by extending this set, i.e. its replenishment with certain "invented" elements ("generalized limits" of divergent sequences, "generalized solutions" of insolvable equations), one can achieve the realization of this procedure (the convergence of any fundamental sequence of rational numbers, the solvability of any additive equation with nonnegative integer parameters).

The described effects need to be rigorously analyzed. Let us consider as an example one of the variants of determining the set of real numbers.

4.4 Real numbers by Cantor

We would like to determine the set of real numbers using the given set of rational numbers $\mathbb{Q}$. One knows that a fundamental sequence of rational numbers can diverge. We would like to show that all real numbers can be interpreted as a special form of limits for the fundamental sequences of rational numbers.

Determine, at first, the very important notions of set theory[8]. Consider a set X and its elements x and y. Denote by $\{x, y\}$ the subset of X that is the elements x and y only. Then $\{y, x\}$ is another denotation of the same subset. Thus, the order of enumeration of elements when writing a set is not important. However, often there is a need to consider an object consisting of two elements x and y, under which the order of enumeration of elements is principled. This object is called the **pair** (x, y). The defining property of the pair is the realization of the equality $(x, y) = (y, x)$ only in the case of equality of the considered elements (see Figure 4.5)[9].

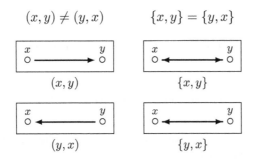

FIGURE **4.5**: Pair and two-element set.

Remark 4.6 In reality, we have used pairs before. Particularly, the metrics and the scalar products are determined on the pairs set of elements of the considered space, see Chapter 3. We shall consider also binary operations, which are determined on the set of pairs too, see Chapter 5.

Definition 4.2 *The set φ of elements pairs of a set X is called the **relation**[10] on X.*

Remark 4.7 More exactly, this is called the binary relation.

We can consider, for example, the relation of the equality for the arbitrary set, the parallelism for the set of lines, the divisibility of the set of natural numbers, the seniority for the set of people, etc. We write $x\varphi y$ if the elements x

and y of the set X belong to the relation φ. Particularly, if φ is the parallelism on the set of lines, then the property $x\varphi y$ is true whenever the line x is parallel to the line y. Analogically, if φ is seniority for the set of people, then the the property $x\varphi y$ is true whenever the person x is older than the person y. Determine now the most important class of relations.

Definition 4.3 *The relation φ on a set X is called the **equivalence**, if the following properties hold (see Figure 4.6) :*

  i) $x\varphi x \ \forall x \in X$ *(the **reflexivity**),*
  ii) *from $x\varphi y$ it follows $y\varphi x \ \forall x, y \in X$ (the **symmetry**),*
  iii) *from $x\varphi y$ and $y\varphi z$ it follows $x\varphi z \ \forall x, y, z \in X$ (the **transitivity**).*

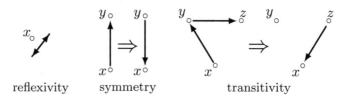

reflexivity   symmetry   transitivity

FIGURE 4.6: Definition of the equivalence.

Remark 4.8 We shall consider another very important relation. This is the order (see Chapter 5).

Particularly, the equality on the arbitrary set, the parallelism on the set of lines, and the similarity of triangles are the equivalences; and the divisibility on the set of natural numbers, seniority on the set of people, and the perpendicularity on the set of lines are not the equivalences.

Example 4.5 *Classification of natural numbers*. Consider the set of natural numbers $\mathbb{N}$. Determine the relation φ there such that the condition $x\varphi y$ is true if the number $|x - y|$ is divisible by three. It is obvious that this relation is equivalence. Now the set $\mathbb{N}$ can be divided by the following subsets

$$N_1 = \{1, 4, 7, ..\}, \ N_2 = \{2, 5, 8, ...\}, \ N_3 = \{3, 6, 9, ...\}.$$

Note that two arbitrary elements from the same of these subsets are equivalent, and two arbitrary elements from the different subsets are not equivalent. Besides, each natural number belongs to the unique subset N_i. $\square$

Remark 4.9 This example is connected with the problem of comparing integers modulo, which deduces the concept of p-adic numbers (see Chapter 6).

Give an extension of Example 4.5.

Definition 4.4 *Let φ be an equivalence on a set X, and x be an element of X. The set $[x]$ of all elements of the set X that are equivalent to x is called the* **equivalence class** *with* **representative** *x.*

Particularly, the set N_1 of Example 4.5 is the equivalence class on the set of natural numbers with representative 1, i.e. $N_1 = [1]$; and N_3 is the equivalence class there with representative 9. Of course, if the elements x and y are equivalent, then their equivalence classes are equal. Therefore, the equivalence class can be determined by its arbitrary element. The equivalence classes for all sets and equivalences satisfy the following properties (see Example 4.5 and Figure 4.7):

i) two arbitrary elements from the same equivalence class are equivalent,

ii) two arbitrary elements from the different subsets are not equivalent,

iii) each element from the given set belongs to the unique equivalence class.

Definition 4.5 *The set of all equivalence classes of the set X with respect to the relation φ is called the* **factor-set** *and is denoted by X/φ.*

The factor-set of the set of natural numbers with respect to the relation φ of Example 4.5 is (see Figure 4.7).

$$\mathbb{N}/\varphi = \{N_1, N_2, N_3\}.$$

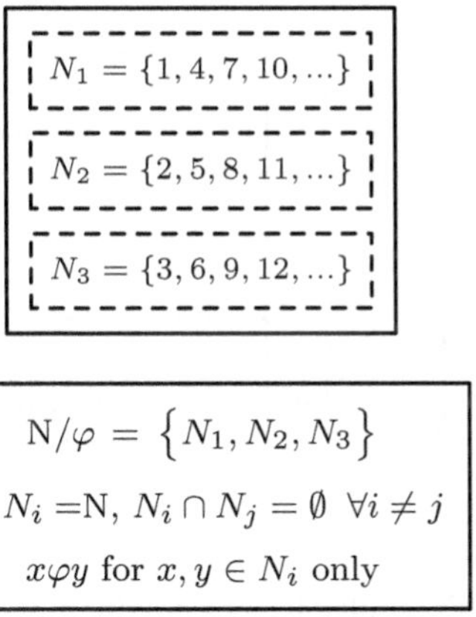

$$N_1 = \{1, 4, 7, 10, ...\}$$

$$N_2 = \{2, 5, 8, 11, ...\}$$

$$N_3 = \{3, 6, 9, 12, ...\}$$

$$\mathrm{N}/\varphi = \{N_1, N_2, N_3\}$$
$$\bigcup N_i = \mathrm{N}, \ N_i \cap N_j = \emptyset \ \forall i \neq j$$
$$x\varphi y \text{ for } x, y \in N_i \text{ only}$$

FIGURE 4.7: Factorization of the set of natural numbers.

Remark 4.10 The factorization is the transformation of a set with elements tantamount to the set of equivalence classes such that the elements of each class have peculiar properties. This is the basis of each classification.

Return to the definition of real numbers. Let F be the set of all fundamental sequences of rational numbers. We determine a relation φ here. Suppose the property $\{x_k\}\varphi\{k\}$ between the sequences $\{x_k\}$ and $\{y_k\}$ of F is true, if for any rational number $\varepsilon > 0$ there exists a number $k(\varepsilon)$ such that $|x_k - y_k| < \varepsilon$ for all $k > k(\varepsilon)$.

Consider the properties of this relation (see Figure 4.8). At first, we have the relation $\{x_k\}\varphi\{x_k\}$ for any sequence $\{x_k\}$ because $|x_k - x_k| = 0$ is the reflexivity. Besides, if we have the relation $\{x_k\}\varphi\{k\}$, then we get $\{y_k\}\varphi\{x_k\}$ because the inequality $|y_k - x_k| < \varepsilon$ is the obvious corollary of the inequality $|x_k - y_k| < \varepsilon$. Therefore, we have the symmetry. Finally, if we have three fundamental sequences $\{x_k\}$, $\{y_k\}$ and $\{z_k\}$ with relations $\{x_k\}\varphi\{k\}$ and $\{y_k\}\varphi\{z_k\}$, then for any $\varepsilon > 0$ there exists a number $k(\varepsilon)$ such that

$$|x_k - y_k| < \varepsilon/2, \quad |y_k - z_k| < \varepsilon/2.$$

Therefore,

$$|x_k - z_k| < |x_k - y_k| + |y_k - z_k| < \varepsilon/2 + \varepsilon/2 = \varepsilon.$$

Then we get $\{x_k\}\varphi\{k\}$. Therefore, the considered relation is the transitivity. Thus, the relation φ on the set F is the equivalence.

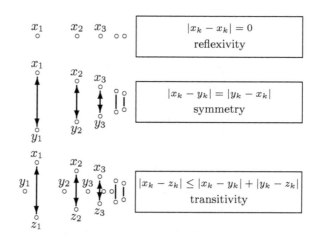

FIGURE 4.8: Equivalence φ of the fundamental sequences.

Now we can consider the definition of the real numbers by Cantor[11].

Definition 4.6 *The factor-set F/φ is called the set of **real numbers** $\mathbb{R}_C$.*

Particularly, the fundamental sequence of rational numbers

$$x_1 = 3, \ x_2 = 3.1, \ x_3 = 3.14, \ x_4 = 3.141, \ x_5 = 3.1415, \dots$$

is the representative of the concrete real number that is denoted by π. Its other representative is the fundamental sequence

$$y_1 = 4, \ y_2 = 3.2, \ y_3 = 3.15, \ y_4 = 3.142, \ y_5 = 3.1416, \ldots$$

that is equivalent to the first sequence. Therefore, we get

$$\pi = \big[\{x_k\}\big] = \big[\{y_k\}\big].$$

We shall denote the equivalence class with the representative $\{x_k\}$ by $[x_k]$ instead $[\{x_k\}]$.

It seems that Definition 4.6 does not quite characterize those objects, which we intuitively associate with real numbers. Indeed, we understand the real numbers as objects similar to rational numbers, and not classes of any sequences. In particular, if the length of both catheti in a right triangle is equal to one (rational number 1), then the length of the hypotenuse turns out to be equal to $\sqrt{2}$, which corresponds to the irrational number. If there is a circle of the unit diameter, then its circumference is equal to π. In these examples, both rational and irrational numbers are the characteristics of single-type objects (lines or curve segments) that are their lengths (see Figure 4.9). Psychologically we are accustomed to perceiving the real numbers as ordinary numbers, similar to rational numbers, but having "weaker" properties in the general case. Therefore, we need to make sure that Definition 4.6 describes a set of real numbers in reality.

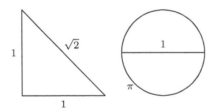

| lengths of the cathetus and the diameter are *rational* |
| lengths of the hypotenuse and the circumference are *irrational* |

FIGURE 4.9: Rational and irrational lengths.

Remark 4.11 The definition of real numbers as classes of equivalent objects does not seem so surprising if you recall that an analogous situation is observed for simpler numerical sets. For example, natural numbers can be determined as classes of equipollent sets (see Chapter 5, Example 5.1), integer numbers can be determined as equivalence classes of pairs of natural numbers (see Chapter 6, Remark 6.8); and rational numbers can be determined as equivalence classes of pairs of integer numbers (see Chapter 6, Definition 6.6), see Table 4.2. Besides, the elements of the very important spaces of integrable functions are equivalence classes of measurable functions (see Chapter 8).

TABLE 4.2:   Sets of numbers as the equivalence classes

set	initial object	construction	equivalence	result	
$\mathbb{N}$	set	set X	X and X' are equipollent	$[X] = [X']$	
$\mathbb{Z}$	natural number or 0	pair (a, b)	$a + b' = a' + b$	$[(a, b)]$	$= [(a', b')]$
$\mathbb{Q}$	integer number	pair (a, b)	$a \cdot b' = a' \cdot b$	$[(a, b)]$	$= [(a', b')]$
$\mathbb{R}$	rational number	sequence $\{x_k\}$	$\{x_k - x'_k\} \to 0$	$[\{x_k\}]$	$= [\{x'_k\}]$

4.5   Conclusions

1. The justification of the determination of mathematical models is based on the passage to the limit.

2. The definition of the limit is not constructive because it uses a priori knowledge of the limit.

3. The proof of the convergence can be based on the Cauchy criterion that uses the fundamentality of the sequences and does not require a priori knowledge of the limit.

4. The Cauchy criterion is not always applicable.

5. The Cauchy criterion is applicable for complete spaces only.

6. The majority of the spaces are non-complete.

7. The classic example of the non-complete spaces is the set of rational numbers.

8. The divergent fundamental sequences of rational numbers determine the irrational numbers by Cantor.

9. The Cantor real numbers are the equivalent classes of the fundamental sequences of rational numbers.

We do not yet have the confidence that a considered set of equivalence classes of fundamental sequences of rational numbers characterizes the real numbers in their natural sense. If this could be proved, we could try to extend this method of determining real numbers to arbitrary metric spaces with divergent fundamental sequences. Perhaps, it is precisely this method that we will be able to use to solve the problem of correctly determining mathematical models of physical processes.

Notes

[1]Examples of non-completeness of metric space are considered in the magnificent book of Gelbaum and Olmsted [57].

[2]The set X with the **anti-discrete topology** has the following property. Each point there has the unique neighbourhood that is the set X. Therefore, any element of any sequence of this space belong to the unique neighbourhood of its arbitrary point. Therefore, each sequence tends to the arbitrary point of the anti-discrete space

[3]The Riemann integral, created by *Bernhard Riemann* in 1854, was the first rigorous definition of the integral of a function on an interval. The Riemann integral and the Riemann integrable functions are considered in the standard courses of mathematical analysis; see, for example, [12], [45], [109], [189].

[4]It is surprising that everywhere the discontinuous Dirichlet function can be obtained as a repeated limit of a sequence of continuous functions

$$D(x) = \lim_{m \to \infty} \lim_{n \to \infty} \cos^{2n}\left(m!\pi x\right),$$

see [45].

[5]The completeness of the metric spaces was determined by *Maurice René Fréchet* in 1906.

[6]The different fixed point theorems are considered, for example, in [3], [5], [87], [88], [95].

[7]The concept of Banach space was proposed by *Stefan Banach, Hans Hahn,* and *Eduard Helly* in 1920. The first results of the Hilbert spaces theory was obtained by *David Hilbert* and *Erhard Schmidt*. The theory of Banach spaces and Hilbert spaces is an important part of functional analysis; see, for example, [42], [71], [81], [91], [95], [100], [123], [142], [147], [203].

[8]The foundations of **set theory** were proposed by *Georg Cantor*, although some results in this direction were obtained earlier by *Bernard Bolzano*. The set theory is the basis of whole mathematics, see [27], [77], [99]. This is a branch of mathematical logic.

[9]It is possible to give the constructive definition of pair (x, y). This is the set with the elements $\{x\}$ and $\{x, y\}$. Indeed, the equality $(x, y) = (y, x)$ here is true, if $x = y$ only

[10]The general concept of relation was proposed by *Augustus de Morgan* in 1847. The relation theory is considered, for example, in [27], [55], [77], [99].

[11]The definition of real number on the basis of fundamental sequences of rational numbers was given by *Georg Cantor* in 1872. Analogical results were obtained also by *Charles Méray* and *Heinrich Eduard Heine*.

Chapter 5

Real numbers and completion

We would like to substantiate the determination of the mathematical models. The general cause of the difficulties here is the passage to the limit in the balance relations for the elementary volume as this volume shrinks to a point. An effective method of the substantiation of the convergence is necessary for this procedure. We know that the standard definitions of the limit are nonconstructive, because the knowledge of the limit is used here. However, we often do not know the fact of the convergence (see Chapter 1 and Chapter 2).

The existence of the limit of the numerical sequences can be determined by the Cauchy criterion (see Chapter 3). This method reduces the convergence of the sequences from its fundamentality. Note that the definition of the fundamental sequence uses only the elements of this sequence that are known. Therefore, one uses the Cauchy criterion for the practical proving of the convergence. This technique is applicable for some functional spaces too, for example, for the space of continuous functions with the standard norm. This circumstance gives us some hope, since in the process of determination of the mathematical physics equations, it is necessary to work with functions, not numbers.

Unfortunately, the Cauchy criterion is not applicable for many metric spaces, particularly, for the sets of rational numbers, positive numbers, continuous functions with integral norm, and Riemann integrable functions. We cannot guarantee the convergence of fundamental sequences there. In reality, the Cauchy criterion is applicable for complete spaces only (see Chapter 4). Moreover, the majority of the mathematical spaces are non-complete. Therefore, we need to have an effective method of passage to the limit for the non-complete spaces too.

Note that for all considered examples of the divergent fundamental sequences there exist points that can be interpreted as "generalized limits" of these sequences. These points are not elements of the given spaces. However, they will be the points of its extensions. Thus, the non-complete spaces can be extended to complete spaces. This idea is realized for Cantor's definition of the set of real numbers. However, we may doubt that this definition really determines the set of real numbers. Besides, it is not clear whether this technique can be used to investigate the convergence of fundamental sequences in an arbitrary metric space.

We consider, at first, the axiomatic definition of the set of real numbers (see Section 5.1). Particularly, each set that satisfies the given list of properties is declared as the set of real numbers. We consider the set of infinite decimals as an example. These are the set of Weierstrass real numbers (see Section 5.2). We prove that this set satisfies all properties of the axiomatic definition of the set of real numbers (see Section 5.3). Therefore, this is an interpretation of the set of real numbers. Then we obtain the analogical result for the set of all equivalence classes of fundamental sequences of rational numbers, i.e. the set of Cantor real numbers (see Section 5.4). The generalization of this result to the general metric spaces is the theorem of completion for the metric spaces (see Section 5.5). This theorem give us the universal method of analysis of the convergence for the general spaces. We shall try to use it to substantiate the determination of mathematical models.

5.1   Axiomatic definition of real numbers

We determined the real numbers as equivalence classes of fundamental sequences of rational numbers. However, it is not clear whether we obtain real numbers, in reality.

Two methods of determining the objects are used in mathematics. In the first case, an object is directly presented in the finished form or constructed on the basis of a constructive algorithm. In the second case, it is asserted that the determined object is all that possesses a specific property. The first method is called **constructive**, and the second method is called **axiomatic**.

Remark 5.1 The constructive approach only is admissible for intuitionistic[1] and constructive[2] mathematics.

It is usually customary to give preference to constructive definitions because the axiomatic method does not allow us to find the object to be determined, but does not even guarantee its existence. Nevertheless, the domain of applicability of the axiomatic approach is much larger. Therefore, the majority of mathematicians recognize this method of definition as admissible. The definition of real numbers by Cantor is constructive. However, the axiomatic definition of the set of real numbers is known too.

Consider a classic example of the axiomatic definition of the numerical set[3].

Example 5.1 *Peano axioms of natural numbers*[4]. The set of **natural numbers** is a set of objects with the following properties

i) There exists a first natural number, which is denoted by 1.

ii) The object n', which is the next after the natural number n, is a natural number too.

iii) The number 1 is not next after any natural number.

iv) If a natural number is next after the natural numbers m and n, then the numbers m and n are equal.

v) If a property is true for the number 1, and from its realization for a natural number, it follows its realization for the next natural number; then this property is true for all natural numbers. □

Now we give the axiomatic definition of the set of real numbers[5].

Definition 5.1 *The set of **real numbers** is the Archimedean field with the Cauchy criterion.*

Remark 5.2 There are other variants of the axiomatic definition of real numbers[6]. Of course, all these definitions are equivalent. However, the Cauchy principle is just most important for us.

It is necessary to verify that the object defined by Definition 4.6 actually satisfies all properties described by Definition 5.1. Therefore, it is necessary to determine, at first, all characteristics of Definition 4.6. We already know what the Cauchy criterion is. This is determined by using a metric. Now we try to determine the definition of the Archimedean field.

The field is defined by operations. This is an algebraic concept of the set with many operations[7]. Determine, at first, a set with unique binary operation that has special properties[8]. The **binary operation** on the set X is a map that transforms two arbitrary elements of X to an element of the same set.

Remark 5.3 We frequently use the denomination "operation" for short, instead of "binary operation".

Definition 5.2 *Let us have a set X and binary operation • here; i.e. for all elements x and y from X an element $x • y$ from X is determined. The set X with operation • is called the **group**[9], if the following properties hold:*

i) $(x • y) • z = x • (y • z)$ *(the **associativity**);*

ii) *there exists an element $e \in X$ such that $x • e = e • x = x$ $\forall x \in X$ (the existence of the **unit**);*

iii) *for any $x \in X$ there exists an element $x^{-1} \in X$ such that $x • x^{-1} = x^{-1} • x = e$ (the existence of the **inverse element**).*

*This group is called **abelian** if it satisfies the following extra condition*

iv) $x • y = y • x$ *(the **commutativity**).*

TABLE 5.1:   Classification of algebraic objects

set	operation	missing property	algebraic class
integer numbers	addition	–	abelian group
non-degenerate matrixes	multiplication	commutativity	group
non-negative integer numbers	addition	existence of the inverse element	monoid
natural numbers	addition	existence of the unit	semigroup
integer numbers	subtraction	associativity	groupoid
natural numbers	subtraction	operation	not a groupoid

The set with a binary operation in the general case is called the **groupoid**; the groupoid with associativity is called the **semigroup**; the semigroup with a unit is called the **monoid** (see Table 5.1).

The set of integer numbers with the operation + (the addition) is the abelian group with the unit 0 and the inverse element $-x$ for any number x. The set of non-degenerate square matrixes (its determinant is not equal to zero) of a concrete order with the matrix multiplication is the group only, because the multiplication of matrixes is not commutative. Addition is the associative operation on the set of non-negative integer numbers; the number zero is the unit here. However, this is not a group, because each of its non-zero elements does not have an inverse element. This is monoid only. Addition is the associative operation on the set of natural numbers too. However, the unit does not exist here. Subtraction is a non-associative operation on the set of integer numbers. Particularly, $(5 - 3) - 1 = 1$, but $5 - (3 - 1) = 3$. Finally, subtraction is not an operation even on the set of natural numbers, because the difference between two natural numbers can be negative.

Now we can give a definition of some of the previously considered classes of mathematical spaces, which is based on the concept of a group.

Definition 5.3 *The abelian group X with operation + (addition) is the **linear space** or the **vector space**[10], if for any element x of X and for any real number a there exists its product ax that is an element of X; besides the following properties holds:*

   i) $a(bx) = (ab)x$ $\forall x \in X$, $a, b \in \mathbb{R}$,

   ii) $1x = x$ $\forall x \in X$,

   iii) $(a + b)x = ax + bx$ $\forall x \in X$, $a, b \in \mathbb{R}$,

   iv) $a(x + y) = ax + ay$ $\forall x, y \in X$, $a \in \mathbb{R}$.

Remark 5.4 It is possible to determine the linear space with the multiplication by the complex numbers.

Definition 5.4 *Let X be the linear space and the topological space too. This is **linear topological space** or **topological vector space**[11], if the addition and multiplication are continuous operators with respect to the topology of the space X.*

Definition 5.5 *The linear space X is called the* **linear normed space** *or the* **normed vector space**[12] *if for any element u of X there exists a non-negative real number $\|u\|$, which is called the* **norm** *of this element such that the following properties hold:*

i) *the norm $\|u\|$ is zero whenever x is the zero element (the unit) of this linear space;*

ii) $\|au\| \leq |a|\|u\| \quad \forall u \in X, \ a \in \mathbb{R}$;

iii) $\|u + v\| \leq \|u\| + \|v\| \quad \forall u, v \in X$.

Definition 5.6 *The linear space X is called the* **unitary space**[13]*, if for any element u, v of X there exists its* **scalar product** (u, v) *that is a real number with the following properties:*

i) $(u, u) \geq 0 \quad \forall u \in X$, *besides,* $(u, u) = 0$ *for the zero element of X only (the non-negativity of the scalar square);*

ii) $(u, v) = (v, u) \quad \forall u, v \in X$ *(the symmetry);*

iii) $\left(a_1 u_1 + a_2 u_2\right) = a_1(u_1, v) + a_2(u_2, v) \ \forall u_1, u_2, v \in X; \ \forall a_1, a_2 \in \mathbb{R}$ *(the linearity with respect to the first multiplier).*

Of course, the scalar product is linear mapping with respect to the second multiplier too, because of the symmetry.

Remark 5.5 It is possible to determine the unitary spaces (and linear normed spaces too) over complex numbers.

The examples of these classes of space were considered before. Now we determine a set with many operations.

Definition 5.7 *Let us have a set X with two binary operations $+$ (addition) and $\cdot$ (multiplication). The set X with these operations is called the* **field**[14] *if the following properties hold*

i) *the set X with addition is the abelian group with unit 0 and the inverse element $-u$ for any $u \in X$;*

ii) *the set X with multiplication is the monoid with unit 1 and the inverse element u^{-1} for any element u of X that is not equal to 0; besides the multiplication is commutative;*

iii) $u \cdot (v + w) = (u \cdot v) + (u \cdot w)$, $\ (v + w) \cdot u = (v \cdot u) + (w \cdot u)$ *for all $u, v, w \in X$ (the* **distributivity***).*

Remark 5.6 We shall consider also the larger class of sets with two operations. This is the ring that is the extension of the field (see Caption 6).

The set $\mathbb{Q}$ of rational numbers is the field.

Remark 5.7 We shall use the field of p-adic numbers in Chapter 6.

By Definition 5.1, the set of real numbers is the Archimedean field. The field is the algebraic object. However, it is necessary to define a non-algebraic notion for determining the Archimedean field.

Definition 5.8 *A set X with relation $\leq$ is called the **ordered set** or **partially ordered set**[15] if this relation is reflexive, transitive, and from $x \leq y$ and $y \leq x$ it follows always $x = y$ (the **antisymmetry**). This relation is called the **order**. The ordered set is called the **linear ordered set**, if for all elements x, y of X one of the following relations holds $x \leq y$ or $y \leq x$.*

Note the following examples of the order: the usual relation $\leq$ for the numerical sets, the divisibility for the natural numbers, and the enclosure for sets. The first of them only is a linear ordered set. Others are not linear ordered sets. The relations $x \leq y$ and $y \geq x$ are equivalent. If $x \leq y$ and $x \neq y$, then we write $x < y$ or $y > x$.

We considered sets with operations and order separately. Now we determine sets with operations and order simultaneously.

Definition 5.9 *Let us have a set X with two binary operations $+$ (addition) and $\cdot$ (multiplication). The set X with these operations is called the **ordered field** if the following properties hold*
1) *if $x \leq y$, then $x + z \leq y + z$ for any $z \leq X$;*
2) *if $0 \leq x$ and $0 \leq y$ then $0 \leq xy$.*

Remark 5.8 The ordered field is the set that is the field and the linear order simultaneously such that its algebraic and ordered properties are consistent. By the way, the distributivity is the consistent property of the given operations on the same set for the field.

The set of rational numbers with standard operations and order is the ordered set.

Definition 5.10 *The ordered field X is the **Archimedean field** if the following **Archimedean axiom**[16] holds: for all elements x, y such that $0 < x < y$ there exists a natural number n such that the sum of n elements x is greater than y.*

Remark 5.9 We shall consider a numerical class without the Archimedean axiom, see Chapter 6.

The set of rational numbers with the standard operations and order is the Archimedean field. We would like to prove that the factor-set $\mathbb{R}_C = F/\varphi$ of Definition 4.6 satisfies all properties of Definition 5.1; i.e. this is in reality the set of real numbers $\mathbb{R}$. However, at first, we consider another definition of real numbers.

Remark 5.10 This is useful, first, because we have to work with different interpretations of the same mathematical object, and second, in order to better understand the relationship between axiomatic and constructive definitions.

5.2   Weierstrass real numbers

We determined the real numbers using the fundamental sequences. However, this numerical class can be determined by another method. Particularly, the real numbers can be interpreted as the infinite decimals. This is the definition of Weierstrass[17].

Definition 5.11 *The **Weierstrass real numbers** are the infinite decimals $x_0, x_1x_2x_3...$, where x_0 is an integer number, and x_1, x_2, x_3,... are decimal digits. Besides, $x_0, x_1x_2...x_{m-1}x_m(9)$ (infinite sequence of digit 9 beginning with $(m+1)$-th symbol after the comma) and $x_0, x_1x_2...x_{m-1}a(0)$ (infinite sequence of digit 0 beginning with $(m+1)$-th symbol after the comma) respectively are the same real number, where $x_m \neq 9$, $a = x_{m+1}$.*

The value before the comma "," of the definition of the real number is called the integer part of the number, and the value after the comma is called the fractional part of the number.

Remark 5.11 One could choose another number system instead of decimal.

Remark 5.12 We shall consider p-adic numbers (see Chapter 6). These objects can be interpreted as infinite fractions, the basis of which is a prime number p, but with an infinite number of digits not after, but before the comma.

The set of all Weierstrass real numbers we denote by $\mathbb{R}_W$. If the values x_0, x_1, x_2,... are equal to zero, then this real number is called zero. This is denoted by 0. The number $x_0, x_1x_2x_3...$ with non-negative x_0 is called a ***positive real number***; this number with negative x_0 is called a ***negative real number***.

Remark 5.13 Obviously, the Weierstrass real number $x_0, x_1x_2x_3...$ can by obtained as the equivalence class of the fundamental sequences (see Caption 4) with representative $\{y_k\}$, where $y_1 = x_0$, $y_2 = x_0, x_1$, $y_3 = x_0, x_1x_2$, $y_4 = x_0, x_1x_2x_3$, etc. The negative Weierstrass real number can be determined analogically. Therefore, the interpretation of the real number by Weierstrass can be reduced to its interpretation by Cantor.

Consider the relations between the Weierstrass real numbers and other numerical sets. Obviously, there exists the bijection between all real numbers $x_0, (0)$ with $x_0 \neq 0$ and the set $\mathbb{N}$ of natural numbers. These real numbers are called ***integer***.

Example 5.1. *Rational numbers.* The rational number, i.e. the element of the set $\mathbb{Q}$, is a fraction p/q, where p is an integer number, and q is a natural number. Consider as the example the rational number $25/4$. Divide the number 25 by 4. The result $6, 25$ can be associated with the Weierstrass real number $6, 25(0)$. Consider now the rational number $25/11$. Dividing the number 25 by 11, we determine integer 2 with an infinite number 27 times after the comma. This result can be associated with the real number $2, (27)$, where the value 27 is repeated an infinite number of times. □

One can prove that each rational number is uniquely representable as an infinite decimal fraction $x_0, x_1x_2...x_m(0)$ or $-x_0, x_1x_2...x_m(0)$ with an infinite number 0 times after position m (we use the short denotation $x_0, x_1x_2...x_m$ or $-x_0, x_1x_2...x_m$ for these numbers) or an infinite decimal fraction, where a natural number is repeated an infinite number of times starting at some position. These real numbers are called **periodic**. The Weierstrass real numbers are called **rational**, if this is a periodic decimal; this is an **irrational** real number, if this is not rational. Obviously, there exists the bijection between the standard rational numbers, i.e. the fractions p/q and the rational Weierstrass real numbers. Note that the rational real numbers only can be non-uniquely represented as an infinite decimal with infinite numbers 0 and 9 times. We will always use the first version of a representation with an infinite number of zeros.

We would like to prove that the set of Weierstrass real numbers satisfies all properties of Definition 5.1. Then it is necessary to determine operations, an order, and a metric on the set $\mathbb{R}_W$.

Determine, at first, the order there. The real numbers x and y are **equal**, if these numbers have the same sign, and all digits of its decimal representations of the same position are equal. The **absolute value** $|x|$ of x is this number, if this is positive or zero. This is the number $x_0, x_1x_2x_3...$, if one considers the negative number $-x_0, x_1x_2x_3...$. For any positive Weierstrass real numbers x and y the relations $x < y$ and $y > x$ are true, if there exists an index $m = 0, 1, 2, ...$ such that $x_m < y_m$, and $x_i = y_i$ for any i less than m. Each positive real number x is greater than zero, i.e. $x > 0$; and each negative number x is less than zero, i.e. $x < 0$. Therefore, each negative real number is less than the arbitrary positive number. Finally, for all negative numbers x and y the relations $x < y$ and $y > x$ are true, if $-y < -x$. Now we determine the **order** on the set of Weierstrass real numbers.

Definition 5.12 *For all numbers* $x, y \in \mathbb{R}_W$ *the relation* $x \le y$ *is true, if* $x = y$ *or* $x < y$.

Remark 5.14 Of course, it is necessary to prove that this is the order in reality. We shall prove it in the next section.

Determine additional notions. Non-empty subset X of $\mathbb{R}_W$ is **upper bounded**, if there exists a number c such that $x \le c$ for any $x \in X$. This is a **lower bounded set**, if there exists a number c such that $c \le x$ for any $x \in X$. The set is called **bounded**, if this is a lower bounded and upper bounded set. The number c is called the **least upper bound** of the numerical set X, if for any $x \in X$ the following inequality holds $x \le c$; and for any number $c_0 < c$ there exists a number $x_0 \in X$ such that $c_0 < x_0$. The least upper bound of the set X is denoted by $\sup X$ or $\sup_{x \in X} X$. The number c is called the **least lower bound** of the numerical set X, if for any $x \in X$ the following inequality holds $c \leq x$; and for all number $c_0 > c$ there exists a number $x_0 \in X$ such that $x_0 < c_0$. The least lower bound of the set X is denoted by $\inf X$ or $\inf\limits_{x \in X} X$. For example, the number 0 is the least upper bound of the set of all negative real numbers and the least lower bound of the set of all positive real numbers. Obviously, the exact upper and lower bound of any set, if it exists, is uniquely determined. Any lower bounded set has a least lower bound, and any upper bounded set has a least upper bound.

Remark 5.15 We shall use these notions in Chapter 7 for the analysis of the optimal control problems.

Determine algebraic operations on the set $\mathbb{R}_W$.

Definition 5.13 *The **sum** of Weierstrass real numbers x and y is the number (see Figure 5.1)*

$$x + y = \sup(a + b),$$

where the least upper bound is determined on the set of all rational numbers x and y such that $a \leq x$ and $b \leq y$.

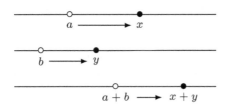

FIGURE 5.1: Addition of real numbers.

Definition 5.14 *The **product** of non-negative Weierstrass real numbers x and y is the number*

$$x \cdot y = \sup(a \cdot b),$$

where the least upper bound is determined on the set of all rational numbers x and y such that $a \leq x$ and $b \leq y$. If the numbers x and y are negative, then $x \cdot y = |x| \cdot |y|$. If we have the product of negative and non-negative numbers, then we have $x \cdot y = -(|x| \cdot |y|)$.

Determine a metric on the set $\mathbb{R}_W$.

Definition 5.15 *The **metric** on the set of Weierstrass real numbers is determined by the equality*

$$\rho(x, y) = |x - y| \quad \forall x, y \in \mathbb{R}_W.$$

Now we prove that the set $\mathbb{R}_W$ satisfies all ordinal, algebraic, and metric properties of Definition 5.1.

5.3 Properties of Weierstrass real numbers

We begin with the order properties[18].

Theorem 5.1 *The set $\mathbb{R}_W$ with relation $\leq$ is the linear ordered set.*

Proof. It is necessary to prove that the relation $\leq$ on the set $\mathbb{R}_W$ is reflexive, transitive, and antisymmetric; i.e. this is the order; besides, this order is linear. The reflexive properties, i.e. the inequality $x \leq x$ for any x, and the comparability of any real numbers x and y that is the linearity of the order follow from the definition of the relation $\leq$.

Suppose now real numbers x and y satisfy both inequalities $x \leq y$ and $y \leq x$. If these numbers are not equal, then we obtain the conditions $x < y$ and $y < x$. However, this case is impossible by the definition of the relation $<$ on the set $\mathbb{R}_W$. Then the equality $x = y$ is true. Therefore, the relation $\leq$ on the set $\mathbb{R}_W$ is antisymmetric.

Now suppose the following inequalities hold $x \leq y, y \leq z$ for all Weierstrass real numbers x, y, and z. If one of these conditions at least is satisfied in the form of equality, then the inequality $x \leq z$ is obvious. Let us have the conditions $x < y, y < z$.

Suppose the number x is positive. Then the numbers y and z are positive too. Consider the representation of the given real numbers as decimals

$$x = x_0, x_1 x_2 ...; \; y = y_0, y_1 y_2 ...; \; z = z_0, z_1 z_2 ... \, .$$

From the inequality $x < y$ the existence of the index m follows such that $x_m < y_m$, and $x_i = y_i$ for any i that is less than m. Analogically, from the inequality $y < z$ the existence of the index n follows such that $y_n < z_n$, and $y_i = z_i$ for any i that is less than n. Denote by l the minimal of the numbers m and n. Then $x_i = y_i = z_i$ for any $i = 0, 1, ..., l-1$. Besides, the following inequalities hold $x_l \leq y_l$ and $y_l \leq z_l$. One of these inequalities is strict. Therefore, we have the strict inequality $x_l < z_l$, because the relation $\leq$ is transitive on the set of natural numbers, and the equality $x_l = z_l$ is impossible in the case of stringency of one of the previous inequalities. Thus, we have $x < z$.

If $x = 0$, then the number y is negative. Using the transitivity of the relation $<$ for the positive real numbers, we obtain $|z| < |x|$. Therefore, we have again $x < z$. Thus, the transitivity of the relation $\leq$ on the set $\mathbb{R}_W$ is proven. This complete the proof of Theorem 5.1. $\square$

Consider additional ordinal properties of real numbers.

Lemma 5.1 *Suppose x and y are real numbers such that $x < y$. Then there exists a rational number a such that $x < a < y$.*

Proof. Consider the numbers

$$x = x_0, x_1 x_2..., \quad y = y_0, y_1 y_2... .$$

Let the number x be positive. Denote by k the minimal index such that $x_k < y_k$. Choose the index $m > k$ such that $x_m < 9$. Then we determine $a = x_0, x_1...x_{m-1} z$, where the digit z is equal to $x_m + 1$. We obtain $x < a < y$. If the numbers x and y have different signs, then we can determine $a = 0$. Finally, if $y \le 0$, then we can find the rational number b such that $|y| < b < |x|$. Now we determine $a = -b$, and the inequality $x < a < y$ is true. $\square$

Lemma 5.2 *For any real number x the following equality holds: $x = \sup a$ with the exact upper bound on the set of all rational numbers such that $a \le x$.*

Indeed, from Lemma 5.1 it follows that for any real number $y < x$ there exists a rational number a such that $y < a \le x$. This complete the proof of Lemma 5.2.

Theorem 5.2 *The set $\mathbb{R}_W$ with the operation $+$ and the order $\le$ is the ordered abelian group.*

Remark 5.16 We determined the ordered field before (see Definition 5.9), where the order relation agreed with the operations of addition and multiplication. Now we have addition only. Thus, the set with group operation $+$ and linear order $\le$ is the **ordered group**, if the inequality $x + z \le y + z$ is true for any element z of this set whenever $x \le y$.

Proof of Theorem 5.2. 1. The sum of all two Weierstrass real numbers is the real number too; i.e. the addition is in reality the operation on the set $\mathbb{R}_W$. The commutativity of the addition follows from the definition of the sum there.

2. Suppose the inequality $x \le y$. For an arbitrary number z find the sum

$$x + z = \sup_{a,c\in\mathbb{Q},\, a\le x,\, c\le z} (a + c).$$

Using the inequality $x \le y$, we get

$$\sup_{a,c\in\mathbb{Q},\, a\le x,\, c\le z} (a + c) \le \sup_{a,c\in\mathbb{Q},\, a\le y,\, c\le z} (a + c) = y + z.$$

Then we obtain $x + z \le y + z$.

From this property, the possibility of the addition of the inequalities follows. Particularly, from the inequalities $x \le y$ and $u \le v$ it follows that

$$x + u \le y + u, \quad y + u \le y + v.$$

By transitivity of the order, we have $x + u \le y + v$.

3. Prove the associativity of the addition. Consider arbitrary real numbers x, y, z. Denote

$$u = (x + y) + z, \quad v = x + (y + z).$$

Suppose the numbers u and v are not equal. For example, let the following inequality $u < v$ be true. Using Lemma 5.1, consider the rational numbers a and b such that

$$u < a < b < v.$$

Choose a natural number n such that

$$b - a > 3 \cdot 10^{-n}. \tag{5.1}$$

Consider rational numbers ξ_n, η_n, ζ_n with n digits after the comma such that the following inequalities hold

$$\xi_n \le x \le \xi_n + 10^{-n}, \quad \eta_n \le y \le \eta_n + 10^{-n}, \quad \zeta_n \le z \le \zeta_n + 10^{-n}.$$

Adding these inequalities, we get

$$\xi_n + \eta_n + \zeta_n \le (x + y) + z \le \xi_n + \eta_n + \zeta_n + 3 \cdot 10^{-n}.$$

Denote

$$r_n = \xi_n + \eta_n + \zeta_n.$$

Then we obtain

$$r_n \le u \le r_n + 3 \cdot 10^{-n}.$$

We have also the analogical inequality

$$r_n \le v \le r_n + 3 \cdot 10^{-n}.$$

Using the properties of the number a and b, we have

$$r_n < a < b < r_n + 3 \cdot 10^{-n}.$$

Therefore, the following inequality holds

$$b - a < 3 \cdot 10^{-n}.$$

However, this contradicts the condition (5.1). Hence, our supposition about non-equality of the numbers u and v is false. Thus, we proved the associativity of the addition on the set $\mathbb{R}_W$.

4. For any number $x \in \mathbb{R}_W$ we have

$$x + 0 = \sup_{a,b \in \mathbb{Q},\, a \le x,\, b \le 0} (a + b) = \sup_{a \in \mathbb{Q},\, a \le x} a.$$

The object at the right-hand side of this equality is x by Lemma 5.2. Using the commutativity of the addition on the set $\mathbb{R}_W$, we prove that the number 0 is, in reality, zero on this set.

5. Now we prove the existence of the inverse element. Consider a real number x and the rational number ξ_n with n digit after the comma such that

$$\xi_n \le x \le \xi_n + 10^{-n}.$$

The number $-x$ that is the number x with the inverse sign satisfies the inequality

$$-\xi_n - 10^{-n} \leq -x \leq -\xi_n.$$

Adding two last inequalities, we get

$$-10^{-n} \leq x + (-x) \leq 10^{-n}.$$

Thus, we have

$$|x + (-x)| \leq 10^{-n}.$$

Suppose the first non-zero digit after the comma for the number $x + (-x)$ is at the m-th position. Then the following inequality holds

$$|x + (-x)| > 10^{-m}.$$

However, this contradicts the previous inequality for $n = m$. Thus, all digits of the decimal $x + (-x)$ are zero. Then $x + (-x) = 0$. $\square$

Theorem 5.3 *The set $\mathbb{R}_W$ with the addition, the multiplication, and the order is the Archimedean field.*

Proof. 1. Obviously, the product of two arbitrary Weierstrass real numbers is the real number too; i.e. the multiplication is, in reality, the operation on the set $\mathbb{R}_W$ too. Its commutativity follows from the definition of the multiplication here.

2. For any number x we have

$$x \cdot 0 = \sup_{a,b\in\mathbb{Q},\, a\leq x,\, b\leq 0} (ab) = \sup_{a\in\mathbb{Q},\, a\leq x} a \cdot 0 = 0.$$

Then $x \cdot 0 = 0 \cdot x = 0$ for any real number x.

3. Consider non-negative real numbers x and y. Prove that their product is non-negative too. We have

$$0 \cdot y = \sup_{a,b\in\mathbb{Q},\, a\leq 0,\, b\leq y} (ab).$$

Using the non-negativity of the number x, we get

$$\sup_{a,b\in\mathbb{Q},\, a\leq 0,\, b\leq y} (ab) \leq \sup_{a,b\in\mathbb{Q},\, a\leq x,\, b\leq y} (ab) = xy.$$

From the two last conditions, it follows that the product xy is non-negative.

4. Prove the possibility of the multiplication of the inequality with non-negative values. Suppose the following inequalities hold

$$x \geq y \geq 0, \quad u \geq v \geq 0.$$

We have

$$yv = \sup_{a,b\in\mathbb{Q},\, a\le y,\, b\le v} (ab) \le \sup_{a,b\in\mathbb{Q},\, a\le x,\, b\le u} (ab) = xu.$$

Thus, $xu \ge yv$.

5. Prove the associativity of the multiplication of real numbers. Consider real numbers x, y, z. If one of them at least is equal to zero, then the associativity property is obvious. In the presence of negative multipliers, the proof of the associativity reduces to analyzing the corresponding absolute values and taking into account the sign of the product. Thus, it suffices to consider the case when all the numbers under consideration are positive.

Denote
$$u = (xy)z, \quad v = x(yz).$$

Consider rational numbers ξ_n, η_n, ζ_n with n digits after the comma such that the following inequalities hold

$$\xi_n \le x \le \xi_n + 10^{-n}, \quad \eta_n \le y \le \eta_n + 10^{-n}, \quad \zeta_n \le z \le \zeta_n + 10^{-n}.$$

Multiplying these inequalities, we get

$$\xi_n\eta_n\zeta_n \le u \le (\xi_n + 10^{-n})(\eta_n + 10^{-n})(\zeta_n + 10^{-n}).$$

The value at the right-hand side of this inequality can be transformed to the sum $\lambda_n + \mu_n$, where $\lambda_n = \xi_n\eta_n\zeta_n$, and the rational number μ_n has the order 10^{-n}. Then we have the inequality

$$\lambda_n \le u \le \lambda_n + \mu_n. \tag{5.2}$$

The condition
$$\lambda_n \le v \le \lambda_n + \mu_n$$

can be obtained analogically. Then we get

$$-\lambda_n - \mu_n \le -v \le -\lambda_n.$$

Adding this result with (5.2), we obtain

$$|u - v| \le \mu_n.$$

Repeating the final part of the associativity of the addition (see Theorem 5.2, step 3), we have $u = v$. Thus, the multiplication on the set $\mathbb{R}_W$ is associative too.

6. Determine the property of the unit. If the number x is non-negative, then we obtain the equalities

$$1 \cdot x = \sup_{a,b\in\mathbb{Q},\, a\le 1,\, b\le x} (ab) = \sup_{b\in\mathbb{Q},\, b\le x} (1\cdot b) = x.$$

Therefore, $1 \cdot x = x = x \cdot 1$ for all x of the set $\mathbb{R}_W$. An analogical result can be obtained for the negative x after the transformation to the corresponding absolute value.

7. Prove the existence of the inverse element with respect to the multiplication for any non-zero real number x. Suppose this number is positive. Consider the number

$$\frac{1}{x} = \sup_{a \in \mathbb{Q},\, a \geq x} \frac{1}{a}. \tag{5.3}$$

The existence of the exact upper bound follows here from the upper boundedness of all numbers $1/a$ because of the condition $a \geq x > 0$. Consider a rational number ξ_n with n digits after the comma such that

$$\xi_n \leq x \leq \xi_n + 10^{-n}. \tag{5.4}$$

From the equality (5.3) it follows that

$$\frac{1}{\xi_n + 10^{-n}} \leq \frac{1}{x}.$$

If $1/\xi_n$ is less than $1/x$, then there exists a number a such that

$$a > x,\ 1/\xi_n < 1/a.$$

Therefore, we get $a < \xi_n$. Using the inequality $\xi_n \leq x$, we conclude that $a < x$. From this contradiction it follows that $1/\xi_n > 1/x$. Hence, we obtain

$$\frac{1}{\xi_n + 10^{-n}} \leq \frac{1}{x} \leq \frac{1}{\xi_n}.$$

Multiply this inequality and (5.4). We have

$$\frac{\xi_n}{\xi_n + 10^{-n}} \leq \frac{1}{x} \cdot x \leq \frac{\xi_n + 10^{-n}}{\xi_n}.$$

Now we get

$$1 - \frac{10^{-n}}{\xi_n + 10^{-n}} \leq \frac{1}{x} \cdot x \leq 1 + \frac{10^{-n}}{\xi_n}.$$

Thus, the following inequality holds

$$\left| \frac{1}{x} \cdot x - 1 \right| \leq \frac{10^{-n}}{\xi_n}.$$

Suppose the m-th digit after the comma of the number x is not equal to zero. Therefore, we have the inequality $10^{-m} \leq \xi_n$ for all $n \geq m$. From the previous inequality, it follows that

$$\left| \frac{1}{x} \cdot x - 1 \right| \leq 10^{-(n+m)}.$$

The number n is arbitrary here. Therefore, we get

$$\frac{1}{x} \cdot x = 1.$$

Thus, the number $1/x$ is inverse to x.

If the number x is negative, then

$$\frac{1}{x} = -\left|\frac{1}{x}\right|.$$

Multiply it by x and use the previous result. We obtain the number 1 of this product. Therefore, the negative real numbers are invertible too with respect to the multiplication.

8. Now prove the distributivity. Consider real numbers x, y, z. If one of them is zero, then this property is obvious. Suppose, at first, that the sum $x + y$ and the number z are positive. Consider again the rational numbers ξ_n, η_n, ζ_n with n digits after the comma such that

$$\xi_n \le x \le \xi_n + 10^{-n}, \quad \eta_n \le y \le \eta_n + 10^{-n}, \quad \zeta_n \le z \le \zeta_n + 10^{-n}.$$

Then we have

$$\xi_n + \eta_n \le x + y \le \xi_n + \eta_n + 2 \cdot 10^{-n}.$$

After multiplication we get

$$(\xi_n + \eta_n)\zeta_n \le (x + y)z \le (\xi_n + \eta_n + 2 \cdot 10^{-n})(\zeta_n + 10^{-n}). \qquad (5.5)$$

Suppose the numbers x and y positive. After multiplication we obtain

$$\xi_n \zeta_n \le xz \le (\xi_n + 10^{-n})(\zeta_n + 10^{-n}), \qquad (5.6)$$

$$\eta_n \zeta_n \le yz \le (\eta_n + 10^{-n})(\zeta_n + 10^{-n}). \qquad (5.7)$$

Adding these inequalities, we have

$$\xi_n \zeta_n + \eta_n \zeta_n \le xz + yz \le (\zeta_n + \eta_n + 2 \cdot 10^{-n})(\xi_n + 10^{-n}). \qquad (5.8)$$

The values at the right-hand sides of the inequalities (5.5) and (5.8) can be transformed to the sum $\lambda_n + \mu_n$, where $\lambda_n = (\xi_n + \eta_n \zeta_n)$, and the rational number μ_n has the order 10^{-n}. Then we get

$$\lambda_n \le (x + y)z \le \lambda_n + \mu_n,$$

$$\lambda_n \le xz + yz \le \lambda_n + \mu_n.$$

From the last inequality, it follows that

$$-(\lambda_n + \mu_n) \le -(xz + yz) \le -\lambda_n.$$

Adding with the previous inequality, we get

$$-\mu_n \leq (x+y)z - (xz+yz) \leq \mu_n.$$

Therefore, we obtain (see the final steps of proving of associativity of the addition and the multiplication)

$$(x+y)z = (xz+yz). \tag{5.9}$$

Suppose the number z is again positive, but one of the numbers x and y, for example, y is negative. Then we can determine the inequalities (5.5) and (5.6), not (5.7). Multiply the inequalities

$$-(\eta_n + 10^{-n}) \leq -y \leq -\eta_n, \ \zeta_n \leq z \leq \zeta_n + 10^{-n}.$$

We have

$$-\eta_n\zeta_n - 10^{-n}\zeta_n \leq -yz \leq -\eta_n\zeta_n - 10^{-n}\eta_n.$$

Add it with the inequality

$$-(\xi_n + 10^{-n})(\zeta_n + 10^{-n}) \leq -xz \leq -\xi_n\zeta_n,$$

which is the corollary of (5.6). We have

$$-\eta_n\zeta_n - \zeta_n 10^{-n} - (\xi_n + 10^{-n})(\zeta_n + 10^{-n}) \leq -(yz+xz) \leq -\eta_n\zeta_n - \eta_n 10^{-n} - \xi_n\zeta_n.$$

Add this inequality with (5.5). We obtain

$$-10^{-n}(\xi_n + 2\zeta_n + 10^{-n}) \leq (x+y)z - (yz+xz) \leq 10^{-n}(\xi_n + 2\zeta_n + 10^{-n}).$$

Applying the known method, we again get the distributivity condition (5.9).

The case of the negative values of the multiplier z can be reduced to the previous case after the transformation to the corresponding absolute value. Thus, the set $\mathbb{R}_W$ with considered operations and order is the ordinal field.

9. Prove that this field is Archimedean. Consider the positive numbers

$$x = x_0, x_1 x_2..., \ y = y_0, y_1 y_2...$$

such that $0 < x < y$. If the number x_0 is non-zero, then we have the inequalities $x_0 \leq x$ and $y \leq y_0 + 1$. Then for the natural numbers x_0 and $y_0 + 1$ there exists a natural number n such that $nx_0 > y_0 + 1$. Now we have

$$y \leq y_0 + 1 < nx_0 \leq nx.$$

Therefore, the Archimedean axiom is realized.

Suppose the first non-zero digit of the decimal representation of the number x is in the m-th position after the comma. Then the following inequality holds $x_m \leq 10^m x$. For the natural numbers x_m, y_0 and the natural number l such that $lx_m > y_0 + 1$. Now we have

$$y \leq y_0 + 1 < lx_m \leq 10^m lx.$$

Finally, determine $n = 10^m l$. This completes the proof of the theorem. □

It remains for us to verify that the Cauchy criterion is true on the set of Weierstrass real numbers. By the definition of convergence for the real numbers (see Caption 3), the sequence of real numbers $\{x_k\}$ tends to a number x, if for any $\varepsilon > 0$ there exists a number $k = k(\varepsilon)$ such that $|x_k - x| < \varepsilon$ for any k that is greater than $k(\varepsilon)$. Determine additional properties of convergence.

Theorem 5.4 (*The principle of nested segments*, see Figure 5.2)[19]. *We consider a sequence of segments such that each successive one of them is embedded in the previous one. If the sequence of lengths of these segments tends to zero, then there exists a unique number belonging to all segments.*

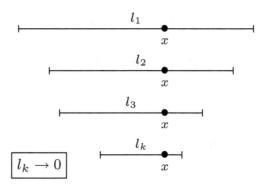

FIGURE 5.2: Principle of nested segments.

Proof. We have a sequence of segments $\{[a_k, b_k]\}$. All segments belong to $[a_1, b_1]$. Then the sequence $\{a_k\}$ of its left ends is bounded by the number b_1. Consider the number $x = \sup a_k$. By definition of the exact upper lower bound, the following inequality holds $a_k \leq x$ for all k. If there exists a number k such that $b_k < x$, then the inequality $b_k < a_l$ for a number l. Then the intervals $[a_k, b_k]$ do not have $[a_l, b_l]$ common points. Then we have the inequality $x \leq b_k$ for all k. Then the point x belongs to all considered segments. Suppose there exists another point y with the same property. Let for definiteness the following inequality $x < y$ be true. From the inequalities $a_k \leq x$ and $y \leq b_k$ it follows that

$$b_k - a_k > y - x > 0, \quad k = 1, 2, \dots.$$

Therefore, the lengths of the intervals do not tend to zero. This contradicts the conditions of the theorem. Thus, the common point of the nested segments is unique. □

One of the most important results of the mathematical analysis is the ***Bolzano–Weierstrass theorem***[20].

Theorem 5.5 *For all bounded sequences of real numbers there exists a convergent subsequence.*

Proof. Consider a bounded sequence $\{x_k\}$. Then all its elements belong to an interval $[a, b]$. Divide this segment in half. One of its halves at least contains infinitely many elements of the given sequence. Denote this half by $[a_1, b_1]$. Let y_1 be an element of this sequence belonging to the last interval (see Figure 5.3). Divide the segment $[a_1, b_1]$ in half. Denote by $[a_2, b_2]$ the half with an infinite set of elements of the sequence $\{x_k\}$. Choose an element of this sequence with number $k_2 > k_1$ belonging to the interval $[a_2, b_2]$ (see Figure 5.3). Then we divide the new segment in half and choose such a half that has infinite elements of the given sequence. Choose an element of this sequence with number $k_3 > k_2$, etc. We obtain the sequence of nested segments $\{[a_k, b_k]\}$; besides, each interval contains the infinite set of elements of the sequence $\{x_k\}$. The lengths of these intervals tends to zero. We have also the numerical sequence $\{y_n\}$ that is a subsequence of $\{x_k\}$ with strictly increasing indexes. By Theorem 5.4, there exists a point x belonging to all segments $[a_k, b_k]$. By the convergence of these lengths to zero, for any $\varepsilon > 0$ there exists a number $k(\varepsilon)$ such that

$$[a_k, b_k] \subset [x - \varepsilon, x + \varepsilon] \ \ \forall k > k(\varepsilon).$$

Using the definition of the numbers y_n, we obtain the inequality

$$\left| y_n - x \right| \leq \varepsilon \ \ \forall k_n > k(\varepsilon).$$

Therefore, we have the convergence $y_n \to x$. Thus, for any bounded sequence of Weierstrass real numbers one can extract a convergent subsequence. $\square$

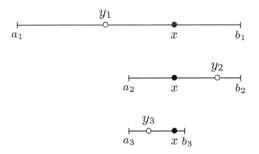

FIGURE 5.3: Proof of the Bolzano–Weierstrass theorem.

Remark 5.17 We shall use the Banach–Alaoglu theorem (see Caption 7 and Caption 9), which is the generalization of the Bolzano–Weierstrass theorem to a large enough class of linear normed space.

Theorem 5.6 *The metric space* $\mathbb{R}_W$ *is complete*[21].

Proof. Consider a fundamental sequence $\{x_k\}$. Therefore, for any value $\varepsilon > 0$ there exists a number $k(\varepsilon)$ such that $|x_l - x_m| < \varepsilon$ for all l and m greater than $k(\varepsilon)$. Particularly, for $\varepsilon = 1$ there exists a number k such that $l, m > k$ the following inequality holds $|x_l - x_m| < 1$. Determine $m = k + 1$. We have

$$|x_l - x_{k+1}| < 1 \ \forall l > k.$$

Then we obtain

$$|x_l| = |x_l - x_{k+1} + x_{k+1}| < 1 + |x_{k+1}|.$$

Denote

$$c = \max\left\{|x_1|, |x_2|, ..., |x_k|, 1 + |x_{k+1}|\right\}.$$

We obtain the inequality $|x_l| < c$ for all large enough l.

Thus, the sequence $\{x_k\}$ is bounded. Using the Bolzano–Weierstrass theorem, we prove the existence of its subsequence $\{y_n\}$ that has a limit x. For any $\varepsilon > 0$ there exist numbers k_1, k_2 such that

$$|x_l - x_m| < \varepsilon/2 \ \forall l, m > k_1,$$

$$y_n < \varepsilon/2 \ \forall n > k_2.$$

Denote $k = \max(k_1, k_2)$, and $n > k$. For all $l > k$ we have

$$|x_l - x| \leq |x_l - y_n| + |y_n - x| \leq \varepsilon/2 + \varepsilon/2 = \varepsilon.$$

Thus, the whole sequence $\{x_k\}$, and not only its subsequence, converges to x. We proved that each fundamental sequence of the set $\mathbb{R}_W$ is convergent. Therefore, this metric space is complete. $\square$

Thus, the set of Weierstrass real numbers satisfies all properties of Definition 5.1. Therefore, we can use the standard denotation $\mathbb{R}$ instead $\mathbb{R}_W$ for this set.

Remark 5.18 There also exists the definition of real numbers by Dedekind as the cuts of rational numbers[22]. The pair (x_-, x_+) of non-empty sets of rational numbers is called the *cut* of the set $\mathbb{Q}$ with lower class x_- and upper class x_+; if they do not intersect, their union coincides with the whole set $\mathbb{Q}$; and any element of the lower class is less than any element of the upper class. For example, the sets

$$x_- = \left\{x \in \mathbb{Q} \mid x \leq 2\right\}, \ x_+ = \left\{x \in \mathbb{Q} \mid x > 2\right\}$$

determine the cut of rational numbers, where the lower class has the largest element 2, and the upper class does not have any smallest elements. The sets

$$x_- = \left\{x \in \mathbb{Q} \mid x \leq 2\right\}, \ x_+ = \left\{x \in \mathbb{Q} \mid x \geq 2\right\}$$

determine the cut, where the lower class does not have any largest elements, and the upper class has the smallest element 2. The cut is determined here by the largest element of the lower class or by the smallest element of the upper class. We declare these two cuts equivalent. However, there exist the cuts with another property. The sets

$$x_- = \{x \in \mathbb{Q} | \, x \le 0 \text{ or } x > 0, \, x^2 < 2\}, \, x_+ = \{x \in \mathbb{Q} | \, x > 0, \, x^2 > 2\}$$

determine the cut of rational numbers too. However, the lower class does not have any largest elements; and the upper class does not have any smallest elements. Therefore, after replacing the strict and non-strict inequalities we obtain the same cut. This is not determined by any rational numbers. Thus, the set of all cuts of rational numbers can be considered as the extension of the set $\mathbb{Q}$.

Now we have the following definition. The set of **Dedekind real numbers** $\mathbb{R}_D$ is the set of all cuts of rational numbers, besides the equivalent cuts that are identified; i.e. this determines the same real number. The rational Dedekind real numbers correspond to the cuts determined by rational numbers, and irrational Dedekind real numbers correspond to the cuts that are not determined by any rational number. Note that it is the rational Dedekind real numbers that are characterized by two equivalent sections, just as rational Weierstrass real numbers can be represented by two types of infinite decimal fractions with an infinite number of zeros or nine.

One proves that the set $\mathbb{R}_D$ with operations of addition and multiplication, order and metric satisfies all properties of Definition 5.1. Therefore, this set is equivalent to the sets of Weierstrass real numbers.

5.4   Properties of Cantor real numbers

Now we can return to the consideration of the factor-set $\mathbb{R}_C = F/\varphi$ of Definition 4.6. Prove that after definition operations and an order there we get in reality the set of real numbers by Definition 5.1.

Determine the addition of the real numbers. Let x and y be real numbers by Definition 4.6. These numbers are determined by fundamental sequences of rational numbers $\{x_k\}$ and $\{y_k\}$. Therefore, we have the equalities $x = [x_k]$, $y = [y_k]$. Consider the sequence of the sums $\{x_k + y_k\}$. We have the inequality

$$\left| (x_m + y_m) - (x_n + y_n) \right| \le |x_m - x_n| + |y_m - y_n|.$$

The sequences $\{x_k\}$ and $\{y_k\}$ are fundamental. Therefore, the terms at the right-hand side of the previous inequality tend to zero as $m, n \to \infty$. Thus, the sequence of sums $\{x_k + y_k\}$ is fundamental too. Then a real number is determined by Definition 4.6. It is called the **sum** $x + y$ of the real numbers and (see Figure 5.4). Thus, for any pair of real numbers we can determine its sum that is a real number too. However, we have a serious question. Does this sum depend on the choice of the fundamental sequences that determine the given real numbers?

Choose now other fundamental sequences of the rational numbers $\{x_k'\}$ and $\{y_k'\}$ that determine the same real numbers x and y. Then we have the inequality

$$\left| (x_k + y_k) - (x_k' + y_k') \right| \le |x_k - x_k'| + |y_k - y_k'|.$$

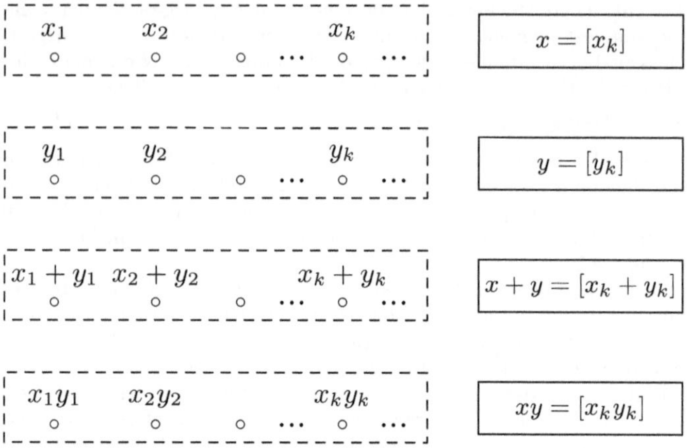

FIGURE 5.4: Operations for real numbers.

The sequences $\{x_k\}$ and $\{x'_k\}$ are equivalent because they determine the same real number, and the sequences $\{y_k\}$ and $\{y'_k\}$ too. Therefore, the value at the right-hand side of the final inequality tends to zero. Hence, the sequences of sums $\{x_k + y_k\}$ and $\{x'_k + y'_k\}$ are equivalent. Then they determine the same real number. Thus, the sum $x + y$ does not depend on the choice of the fundamental sequences that determine the summands. This is determined by the whole equivalence classes that are the summands. Therefore, the sum is determined unequivocally.

Lemma 5.3 *The set* $\mathbb{R}_C$ *with operation* $+$ *is the abelian group.*

Proof. Consider real numbers $x = [x_k]$, $y = [y_k]$. Using the definition of the addition here, we have the equalities

$$x + y = [x_k + y_k] = [y_k + x_k] = y + x,$$

because of the commutativity of the addition for the rational numbers. Thus, the addition of the real numbers is commutative. Now consider three real numbers $x = [x_k]$, $y = [y_k]$, and $z = [z_k]$. We get

$$(x + y) + z = \big[(x_k + y_k) + z_k\big] = \big[x_k + (y_k + z_k)\big] = x + (y + z),$$

because of the associativity of the addition for the rational numbers. Therefore, the addition of the real numbers is associative too.

Determine the real number $\theta = [r_k]$ that is determined by a sequence of rational numbers $\{r_k\}$ with zero as the limit. Then for any number $x = [x_k]$ we have

$$x + \theta = [x_k + r_k] = [x_k] = x,$$

because the sequence $\{r_k\}$ tends to zero. Therefore, the number θ is the zero element on the set $\mathbb{R}_C$ with respect to the given addition. We denote it by 0.

Now for any real number $x = [x_k]$ we determine the number $-x$ that is equal to $[-x_k]$. We have

$$x + (-x) = \left[x_k + (-x_k)\right] = [r_k] = 0,$$

because the sequence with zero elements only is equal to $\{r_k\}$. Hence, the number $-x$ is inverse to x with respect to the addition. Thus, the set of real numbers that is determined by Definition 4.6 is the abelian group. □

Analogically, we can determine the product xy of the real numbers x and y as the real number that is determined by the sequence of product $\{x_k y_k\}$ (see Figure 5.4), where the fundamental sequences of rational numbers $\{x_k\}$ and $\{_k\}$ determine the given real numbers. Of course, it is necessary to prove that the sequence of the product is fundamental, and the result does not depend on the fundamental sequences that determine the numbers x and y. We can prove also that this multiplication of real numbers is commutative and associative. Besides, there exists a unit with respect to the multiplication that can be denoted by 1. Then for any non-zero real number there exists an inverse number with respect to the multiplication. Finally, the following distributive condition holds $x(y + z) = xy + xz$. All these properties are substantiated by a known technique. Thus, we have the following result.

Lemma 5.4 *The set $\mathbb{R}_C$ with addition and multiplication is the field.*

We determine now the order on the set of real numbers. The fundamental sequence of rational numbers is called positive, if there exists a positive rational number r such that all elements of this sequence with large enough numbers are greater than r. If the fundamental sequence is positive, then each equivalent fundamental sequence is positive too. Therefore, the positivity is the same property of all equivalent fundamental sequences. Then we say that the real number is positive, if it is determined by positive fundamental sequences of rational numbers (see Figure 5.5). The relation $x < y$ for real numbers x and y is true, if the difference $y - x$ is positive. Let the relation $x \leq y$ be true if $x < y$ or $x = y$.

Lemma 5.5 *The set $\mathbb{R}_C$ with relation $\leq$ is the linear ordered set.*

Proof. Of course, $x \leq x$ because of the equality $x = x$. Suppose the condition $x \leq y$ and $y \leq z$, where $x = [x_k]$, $y = [y_k]$, and $z = [z_k]$. Then there exist positive rational numbers r and s such that $y_k - x_k > r$ for any large enough number k or $x = y$ and $z_k - y_k > s$ for any large enough number k or $y = z$. If $x = y$ or $y = z$, then the inequality $x \leq z$ is obvious. Suppose we have both inequalities. Then we have

$$z_k - x_k = (z_k - y_k) + (y_k - x_k) > s + r,$$

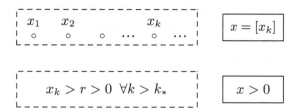

FIGURE 5.5: Positivity of a real number.

and the inequality $x \leq z$ is true. Hence, the relation $\leq$ is transitive.

Suppose now the following inequalities hold: $x \leq y$ and $y \leq x$, where $x = [x_k]$, $y = [y_k]$. If we do not have the equality $x = y$, then the strong inequalities $x < y$ and $y < x$ are true. Using the first inequality, we determine the existence of the positive number r such that $y_k - x_k > r$ for any large enough number k. Using the second inequality, we determine the existence of the positive number s such that $x_k - y_k > s$ for any large enough number k. Then we have both inequalities $y_k > x_k$ and $x_k > y_k$ for any large enough number k that is impossible. Therefore, the equality $x = y$ is true. Thus, the relation $\leq$ on the set of real numbers is the order in reality.

Prove the linearity of the order. Consider arbitrary elements $x = [x_k]$, $y = [y_k]$. If the sequences $\{x_k\}$ and $\{y_k\}$ are equivalent, then we have the equality $x = y$. Suppose now these sequences are not equivalent. Then there exists a number n such that we have the inequality $x_k \leq y_k$ or $y_k \leq x_k$ for all $k > n$ because these sequences are fundamental. If the first inequality is true, then we have $x \leq y$ by the definition of the order. Analogically, from the second inequality, it follows that $y \leq x$. Thus, we have the linear ordered set. $\square$

We proved the algebraic and ordered properties separately. Now we determine the following result.

Lemma 5.6 *The set $\mathbb{R}_C$ is the Archimedean field.*

Proof. Suppose for any elements $x = [x_k]$, $y = [y_k]$ of $\mathbb{R}_C$ we have the inequality $x \leq y$. Therefore, $x_k \leq y_k$ for any large enough number k. Then for all $z = [z_k]$ we have $x + z = [x_k + z_k]$, $y + z = [y_k + z_k]$, and $x_k + z_k \leq y_k + z_k$ for large enough k. Therefore, we get $(+ z) \leq (+ z)$. Suppose $0 \leq x$ and $0 \leq y$. Then we have the inequality $0 \leq x_k$ and $0 \leq y_k$ for large enough k. Hence, $0 \leq x_k y_k$ for large enough k. Therefore, $0 \leq xy$. Thus, the set $\mathbb{R}_C$ is the ordered field.

Prove that the Archimedean axiom is true too. Consider an element $z = [z_k]$ of $\mathbb{R}$. The sequence $\{z_k\}$ is fundamental. Therefore, this is the bounded sequence. Then there exists a natural number m such that $z_k < m$ for all k. The sequence $\{m - z_k\}$ is fundamental and positive. Then we have $z < n$, where n is the real number determined by the stationary sequence with element m. Consider elements x and y such that $0 < x < y$. Therefore, for the real number $y/x = yx^{-1}$ there exists a natural number n such that $y/x < n$. Now we have $nx > y$. Thus, the Archimedean axiom is true, and the set $\mathbb{R}_C$ is the Archimedean field. $\square$

Determine now metric properties of the set of real numbers. For real numbers $x = [x_k]$ and $y = [y_k]$ consider the sequence $\{|x_k y_k|\}$. We have the quadrangle inequalities (see Figure 5.6)

$$|x_k - y_k| \le |x_k - x_m| + |x_m - y_m| + |y_m - y_k|,$$

$$|x_m - y_m| \le |x_m - x_k| + |x_k - y_k| + |y_k - y_m|.$$

Therefore, we get

$$\big||x_k - y_k| - |x_m - y_m|\big| \le |x_k - x_m| + |y_m - y_k|.$$

The sequences $\{x_k\}$ and $\{y_k\}$ are fundamental. Hence, the terms at this right-hand side of the final inequality tend to zero. Then the rational sequence $\{|x_k - y_k|\}$ is fundamental. Therefore, it determines a real number d.

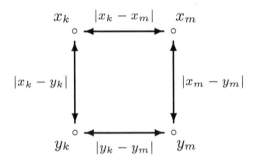

FIGURE 5.6: Quadrangle inequality.

Prove that the value d does not depend on the choice of the fundamental sequences. Choose other fundamental sequences $\{x'_k\}$ and $\{y'_k\}$ that determine the numbers x and y. We have the analogical relations

$$|x_k - y_k| \le |x_k - x'_k| + |x'_k - y'_k| + |y'_k - y_k|,$$

$$|x'_k - y'_k| \le |x'_k - x_k| + |x_k - y_k| + |y_k - y'_k|.$$

Therefore, we get the inequality

$$\big||x_k - y_k| - |x'_k - y'_k|\big| \le |x_k - x'_k| + |y_k - y'_k|.$$

Hence, the sequences $\{|x_k - y_k|\}$ and $\{|x'_k - y'_k|\}$ are equivalent. Therefore, they determine the same real number. Thus, the number d does not depend on the choice of fundamental sequences that determine the numbers x and y. Hence, for all these numbers we can determine the concrete real number $d = d(x, y)$ (see Figure 5.7).

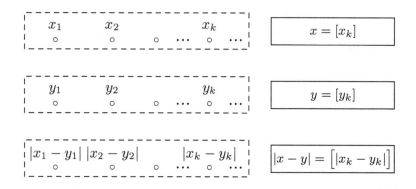

FIGURE 5.7: Metric on the set of real numbers.

Lemma 5.7 *The set $\mathbb{R}_C$ with operator $d : (\mathbb{R}_C)^2 \to \mathbb{R}_C$ is the metric space.*

Proof. The values of d are non-negative values only, because these real numbers are determined by positive fundamental sequences of rational numbers. Besides, it is true, if for the corresponding fundamental sequences $\{x_k\}$ and $\{y_k\}$ the sequence $\{|x_k - y_k|\}$ tends to zero only. Therefore, it determines the same real number. Thus, the equality $d(x, y) = 0$ is true if and only if $x = y$. The equality $d(x, y) = d(y, x)$ is the corollary of the formula $|x_k - y_k| = |y_k - x_k|$. Now we have

$$d(x, y) = \lim |x_k - y_k| \le \lim |x_k - z_k| + \lim |z_k - y_k| = d(x, z) + d(z, y).$$

This is the triangle inequality. Therefore, d is the metric of the set of Cantor real numbers. $\square$

Denote the value $d(x, y)$ by $x - y$. Then the sequence $\{x_k\}$ of real numbers tends to a real number x, if $d(x_k, x) \to 0$, that is $|x_k - x| \to 0$. This is the standard definition of the convergence on the space of real numbers. Note that the metric structure of the set of real numbers is consensual with its algebraic and ordered structures. Particularly, if we have the convergence $x_k \to x$ and $y_k \to y$, then $x_k + y_k \to x + y$ and $x_k y_k \to xy$. Besides, from the inequality $x_k \le y_k$ for all k it follows that $x \le y$.

Our next step is the interpretation of the rational numbers as elements of the determined set $\mathbb{R}$. We know that fundamental sequences of rational numbers can diverge. However, there exist fundamental sequences with limit. Therefore, the arbitrary equivalent fundamental sequence has the same limit. The corresponding equivalence class determines a concrete real number by Definition 4.6. Each rational number is the limit of a rational sequence. We can choose as this sequence the stationary one with elements equal to this rational number. Therefore, there exists the bijection between the set $\mathbb{Q}$ of all rational numbers and the subset $\mathbb{Q}'$ of the set of real numbers that are determined by convergent fundamental sequences (see Figure 5.8).

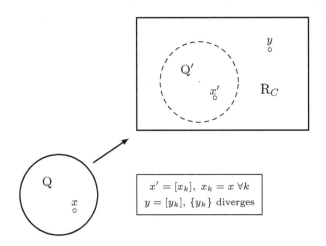

FIGURE 5.8: Set of rational numbers is isomorphic to the subset $\mathbb{Q}'$ of $\mathbb{R}_C$.

For any rational number x there exists an element x' of the set $\mathbb{Q}'$. The stationary sequence with elements equal to x is its representative. On the contrary each element x' of $\mathbb{Q}'$ has convergent fundamental sequences as representative with a same limit x. Each algebraic, topological, etc. property of the set of rational numbers has an analogue on the set $\mathbb{Q}'$. In this situation the set $\mathbb{Q}$ and $\mathbb{Q}'$ are called *isomorphic*.

Now consider the very important problem of the approximation of real numbers.

Lemma 5.8 *Each real number can be approximated by rational numbers.*

Proof. Let x be a real number that is determined by a sequence $\{x_k\}$ of rational numbers. Denote by y_k the real number that is determined by the stationary sequence with element x_k. For any number n the difference $x - y_n$ is determined by the sequence $\{x_k - x_n\}$. Therefore, the value $|x_k - x_n|$ is small enough for large enough numbers k and n by the definition of the fundamental sequence. Hence, the sequence $\{x - y_n\}$ tends to zero as $n \to \infty$. Then we have the convergence $y_n \to x$. Using the isomorphism between the rational numbers x_k and the rational real numbers y_k, we prove that the arbitrary real number x is the limit of a sequence of rational numbers. Therefore, it can be approximated by rational numbers with arbitrary exactness. □

Remark 5.19 More exactly, the set of rational numbers is dense in the space of real numbers by Lemma 5.8. We shall give the exact definition of the density soon.

The last result is extremely important. This is the basis of practical use of the irrational numbers. Indeed, how we can use the irrational numbers, for example, the number π, in calculations? The axiomatic definition is not applicable, because of its non-constructiveness. The interpretation of real numbers as cuts of the set of rational numbers (Dedekind definition) does not use the real information for practical calculation of the irrational number. We can apply infinite decimal fractions (Weierstrass definition), if we know the necessary quantity of digits of this representation. However, the Cantor definition guarantees the possibility of the approximation of the arbitrary real number by rational numbers. For example, the number π is the sum of the concrete series with rational components. We could use the partial sum of this series for obtaining the rational approximation of π with arbitrary exactness.

Now prove the completeness of the set of real numbers.

Lemma 5.9 *The metric space $\mathbb{R}_C$ is complete.*

Proof. Let $\{y_k\}$ be a fundamental sequence of real numbers. Consider a sequence $\{\varepsilon_k\}$ of rational numbers with zero limit. Using the possibility of the approximation of the arbitrary real number by a rational number, we obtain the existence of a rational number x_k such that $|y_k - x_k| \le \varepsilon_k$. Therefore, we get the inequality

$$|x_m - x_n| \le |x_m - y_m| + |y_m - y_n| + |y_n - x_n| \le \varepsilon_m + |y_m - y_n| + \varepsilon_n.$$

The sequence $\{y_k\}$ is fundamental. Then the value at the right-hand side of the last inequality tends to zero as $m, n \to \infty$. Hence, the sequence $\{x_k\}$ of rational numbers is fundamental. Therefore, it determines a real number y. We have

$$|y_k - y| \le |y_k - x_k| + |x_k - x|.$$

The first term at the right-hand side of this inequality is small enough for the large enough value k because the rational number $_k$ approximates the real number y_k. The second term tends to zero here because the real number y is determined by the sequence $\{y_k\}$. Therefore, the arbitrary fundamental sequence of real numbers converges. Hence, the set of real numbers is complete. □

Remark 5.20 We extended the set of rational numbers $\mathbb{Q}$ to the complete set of real number $\mathbb{R}$ such that the set $\mathbb{Q}$ (accurate within isomorphism) is dense to $\mathbb{R}$. In this situation, the set of the real number is called the ***completion*** of the set of the rational number.

We proved the following result.

Theorem 5.7 *The set $\mathbb{R}_C$ that is determined by Definition 4.6 is the set of real numbers by Definition 5.1.*

Thus, we can use the standard denotation $\mathbb{R}$ instead $\mathbb{R}_C$ for the set of Cantor real numbers.

Remark 5.21 All constructive definitions of real numbers are equivalent. We have actually the same set of $\mathbb{R}$ in different interpretations in all cases.

Remark 5.22 The non-uniqueness of the real number interpretations is not so surprising. Other axiomatic definitions have analogical properties. For example, the set $\{1, 2, 3, ...\}$ is not a unique interpretation of the axiomatic definition of the set of natural numbers (see Example 5.1). Particularly, the set $\{5, 6, 7, ..\}$ satisfies the Peano axioms with first element 5. The set $\{0, 1, 2, \}$ with first element 0 is often called the set of natural numbers. One can prove that the set of numbers $2n$ with arbitrary natural n satisfies the Peano axioms too. The number 2 is the first element here. It would seem that there are completely different objects in these interpretations. However, all these sets have exactly the same properties within the framework of the theory of natural numbers. Since in mathematics, objects themselves are not studied, but their properties are. The different interpretations of the concrete axiomatic definition are equivalent.

Now we would like to extend Cantor's idea to the general metric spaces.

5.5 Completion of metric spaces

What have we just proved? The space of rational numbers is non-complete. However, it can be extended such that this extension is the complete space. Besides each element of this extension can be approximated by elements of the initial spaces. We would like to obtain the analogical result for the general metric space. Let X be a metric space with metric ρ.

Definition 5.16 *The subset   of   is **dense**, if each element of X is the limit of a sequence of M.*

For example, the set of all rational numbers is dense in the space of real numbers. If the set M is dense in the metric space X, then each element of X can be approximated by elements of M (see Lemma 5.8).

Definition 5.17 *The space X is **isometric** to the space Y with metric d, if there exists a bijection $A : X \to Y$ such that the distance between all points after the mapping A and its inverse mapping does not change (see Figure 5.9).*

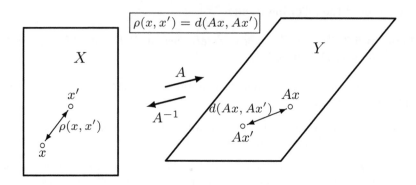

FIGURE 5.9: Spaces X and Y are isometric.

If the spaces X and Y are isometric, then the following equalities hold

$$d(Ax, Ax') = \rho(x, x') \ \forall x, x' \in X,$$

$$\rho(y, y') = d(A^{-1}y, A^{-1}y') \ \forall y, y' \in Y.$$

If two spaces are isometric, then they have the same metric properties.

Remark 5.23 Isometrics is a partial case of the *isomorphism*, which is realized for the metric spaces. Two isomorphic objects of a mathematical theory have the same properties with respect to this theory. Thus, within the framework of this theory, it is not possible to distinguish isomorphic objects, so that they can be considered as one and the same object. In particular, the various interpretations of an axiomatic definition are isomorphic in the sense of a theory corresponding to the given definition. In this connection, it can be said that particular mathematical theories study objects to within an isomorphism of the corresponding theory. In Chapter 6 we shall consider isomorphism of rings.

Definition 5.18 *The complete metric space Y is called the* **completion** *of the space X, if X is isometric to a dense subset of Y.*

Particularly, the space of real numbers is the completion of the space of rational numbers by Theorem 5.7.

Note also the **quadrangle inequality** that is the corollary of the triangle inequality (see Figure 5.10)

$$\rho(a, d) \leq \rho(a, b) + \rho(b, c) + \rho(c, d) \ \forall a, b, c, d \in X.$$

Our general result is the **completion theorem**[23].

Theorem 5.8 *For all metric space, there exists its completion.*

Proof. 1. Let X be a metric space with metric ρ. Consider the set F of all its fundamental sequences. Determine the relation φ on F. Suppose the condition $\{x_k\}\varphi\{y_k\}$ on the set F is true if the sequence $\{\rho(x_k, y_k)\}$ tends

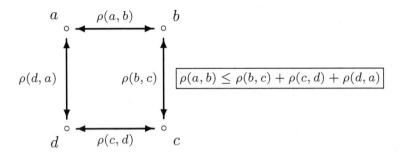

FIGURE 5.10: Quadrangle inequality.

to zero as $k \to \infty$. It is obvious that φ is the equivalence. The factor-space $Y = F/\varphi$ is the analogue of the set $\mathbb{R}$ that is determined by Definition 4.6.

2. For all points x, y of the set Y choose sequences $\{x_k\}$ and $\{y_k\}$ such that $\{x_k\}$ belongs to the equivalence class x, and $\{y_k\}$ belongs to the equivalence class y, that is $x = [x_k]$, $y = [y_k]$. Using the quadrangle inequality, we get

$$\rho(x_m, y_m) - \rho(x_n, y_n) \leq \rho(x_m, x_n) + \rho(y_m, y_n),$$

$$\rho(x_n, y_n) - \rho(x_m, y_m) \leq \rho(y_m, y_n) + \rho(x_m, x_n).$$

Hence, we have the inequality

$$\left| \rho(x_m, y_m) - \rho(x_n, y_n) \right| \leq \rho(x_m, x_n) + \rho(y_m, y_n).$$

The terms at the right-hand side of this formula tend to zero, because the sequences $\{x_k\}$ and $\{y_k\}$ are fundamental. Therefore, the sequence of real numbers $\{\rho(x_k, y_k)\}$ is fundamental too. Then it has a limit because of the completeness of the space $\mathbb{R}$. Determine the value

$$d = d(x, y) = \lim_{k \to \infty} \rho(x_k, y_k).$$

This limit depends, in principle, on the choice of the concrete fundamental sequences. Now let the sequence $\{u_k\}$ be equivalent to $\{x_k\}$, and $\{v_k\}$ be equivalent to $\{y_k\}$. Using the previous technique, obtain the inequality

$$\left| \rho(u_k, v_k) - \rho(x_k, y_v) \right| \leq \rho(x_k, u_k) + \rho(y_k, v_k).$$

Passing to the limit, we obtain the equality of the limits of the sequences $\{\rho(u_k, v_k)\}$ and $\{\rho(x_k, y_k)\}$. Therefore, these sequences determine the same real number, that is the value $d(x, y)$ does not depend on the concrete sequences that define the classes x and y. It is determined by the elements x and y only. Determine its properties.

The number $d(x, y)$ is non-negative as the limit of the sequence with non-negative elements. Besides, for any element of Y we get

$$d(x, x) = \lim_{k \to \infty} \rho(x_k, x_k) = 0.$$

If the value $d(x, y)$ is zero, then the fundamental sequences of and are equivalent. Therefore, we have the equality $x = y$. Thus, the condition $d(x, y) = 0$ is true if and only if the elements x and y are equal. The symmetry of the functional d follows from the analogue property of the metric ρ. Besides, for all elements x, y and z with corresponding sequences $\{x_k\}$, $\{y_k\}$ and $\{z_k\}$, we get

$$d(x, z) = \lim_{k \to \infty} \rho(x_k, z_k) \leq$$

$$\lim_{k \to \infty} \rho(x_k, y_k) + \lim_{k \to \infty} \rho(y_k, z_k) = d(x, y) + d(y, z).$$

Thus, the functional d is metric. This is the analogue of the metric on the set of real numbers. Therefore, the pair (Y, d) is the metric space.

3. Prove that our initial space X can be interpreted as a part of the obtained space. For all $x \in X$ determine the stationary sequence with element x. It obvious that it is fundamental. Then it determines a class equivalence x'. Determine the operator $A : X \to Y$ such that $Ax = x'$. Define the image $X' = A(X)$ (see Figure 5.11). The elements of X' are determined by converged fundamental sequences only. Note that each element x' of X' from is determined by the concrete element x. Indeed, the concrete element x' is determined by the convergent sequences of X with the same limit. This limit is the element x that corresponds to x'. Therefore, the operator $A : X \to X'$ is invertible.

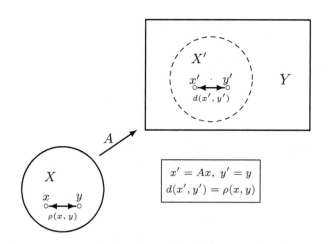

FIGURE 5.11: Isometry of the spaces X and X'.

For all elements x, y of X determine the value

$$d(Ax, Ay) = \lim_{k \to \infty} \rho(x_k, y_k).$$

Let these converged sequences be stationary here. Then we have the equality $d(Ax, Ay) = \rho(x, y)$. For all x', y' of X' the number $d(x', y')$ is the limit of the numerical sequence $\{\rho(x_k, y_k)\}$. We have the elements of the set X'. Therefore, we can choose the stationary sequences with elements $x = A^{-1}x'$ and $y = A^{-1}y'$ as the sequences $\{x_k\}$ and $\{x_k\}$. Then we obtain the equality

$$\rho(A^{-1}x', A^{-1}y') = d(x', y').$$

Thus, the metric space X is isometric to the subspace X' of Y (see Figure 5.11).

The isometric sets X and X' are isomorphic by the metric spaces theory. Hence, we can interpret the initial set X as a subset of Y, and the set Y as an extension of the set X. Besides, the elements of the initial set can be identified with corresponding equivalence classes. Then we can interpret it as elements of the set Y. Particularly, the rational numbers can be interpreted as real numbers.

4. Let be an arbitrary element of the set Y. It is determined by a fundamental sequence $\{x_k\}$. Using a definition of the metric d, we have

$$D(Ax_k, x) = \lim_{n \to \infty} \rho(x_k, x_n),$$

because the equivalence class Ax_k is determined by the stationary sequence with element x_k, and is determined by the fundamental sequence $\{x_n\}$.

For all $\varepsilon > 0$ we can choose the numbers k, n large enough that the inequality $\rho(x_k, x_n) < \varepsilon$ holds. Using the last equality, we have the convergence $Ax_k \to x$ in Y. Identify the element x_k with equivalence class Ax_k; we get the convergence $x_k \to x$ in Y. Thus, the set X (in reality, its isometric image by the operator A) is the dense subset of Y. Therefore, each element of the extended set can be approximated by elements of the initial set. Particularly, each real number can be approximated by a rational one.

5. Consider a fundamental sequence $\{y_k\}$ of the space Y. Let the numerical sequence $\{\varepsilon_k\}$ tend to zero. Using the density of the inclusion of the set X to Y, we obtain the existence of an element x_k with large enough value k that $d(x_k, y_k) < \varepsilon_k$. Then we have (see Figure 5.12)

$$\rho(x_m, x_n) = d(Ax_m, Ax_n) \leq d(Ax_m, y_m) + d(y_m, y, n) + d(y_n, Ax_n) <$$

$$\varepsilon_m + d(y_m, y, n) + \varepsilon_n$$

by the quadrangle inequality. The value at the right side of this inequality tends to zero because of the fundamentality of $\{y_k\}$ and the properties of the sequence ε_k. Therefore, the sequence $\{x_k\}$ is fundamental with respect to the space X. Therefore there exists an element y of Y such that $Ax_k \to y$ in Y.

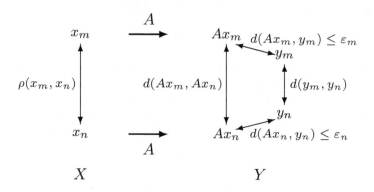

FIGURE 5.12: Proof of the completeness.

We have the inequality

$$d(y_k, y) \le d(y_k, ax_k) + d(Ax_k, y) < \varepsilon_k + d(Ax_k, y).$$

Then we get $y_k \to y$ in Y. Thus, the arbitrary fundamental sequence of the space Y has a limit. Therefore, this space is complete. Thus, the set X is isometric to the dense subset of the complete metric space Y. Hence, each metric space has a completion. This completes the proof of Theorem 5.8. □

Remark 5.24 The analogue of the completion theorem is true for uniform spaces too[24].

The proof of this theorem gives the technique for practical completion of metric spaces.

Example 5.2 *Space of positive numbers.* Consider the space X of positive numbers with standard metric $\rho(x, y) = |x - y|$ for all x and y from X. Find its completion. Determine the set F of all fundamental sequences on the set X. The sequences $\{x_k\}$ and $\{y_k\}$ are in the relation φ if $|x_k - y_k| \to 0$ as $k \to \infty$. Consider the factor-set $Y = F/\varphi$. This is the union of the set X' that is determine be the convergent fundamental sequences from X and the set Y' that is determined to be the divergent fundamental sequences from X. Note that each divergent fundamental sequence from as is equivalent to the sequence $\{\theta_k\}$, where $\theta_k = 1/k$, $k = 1, 2, \ldots$. Therefore, the set Y' consists of the unique element $\theta = \theta_k$. Each element of X' is determined by the stationary fundamental sequence of X, namely by the concrete element of X. Determine the properties of the element θ. For any element $x = [x_k]$, $y = [y_k]$ we can determine its sum $x + y = [x_k + y_k]$. Particularly, we have $x + \theta = [x_k + \theta_k]$. Note, that $(x_k + \theta_k) + \theta_k$. This value tends to zero. Therefore, the sequences $\{x_k + \theta_k\}$ and $\{x_k\}$ are equivalent. Hence, they determine the same element of the set Y. Thus, we have the equality $x + \theta = x$ for any element x from Y. Therefore, the element θ can be interpreted as zero. □

TABLE 5.2: Completion of metric spaces

non-complete space	completion
rational numbers	non-negative numbers
continuous functions with integral metric	Lebesgue integrable functions
Riemann integrable functions	Lebesgue integrable functions

The results of using the completion theorem for the considered examples are given at Table 5.2. Particularly the completion of the set of rational numbers is the set of real numbers. Consider the set X of positive numbers with standard metric. There exist fundamental sequences $\{x_k\}$ here such that for all $\varepsilon > 0$ there exists a number $k(\varepsilon)$ such that the inequality $|x_k| < \varepsilon$ is true for all k greater than $k(\varepsilon)$. These sequences are equivalent. The relevant equivalence class is the number zero. Therefore, the completion of the set of positive numbers is the set of non-negative numbers. We can prove also that the completion of the set of continuous functions and Riemann integrable functions with integral metric is the set of Lebesgue integrable functions (see Chapter 8).

Thus, for any fundamental sequence of the arbitrary metric space we can have two different cases only. Maybe this sequence converges. Then its limit is a point of the initial space. However, maybe the fundamental sequence does not have any limits on the initial space. Then we can determine the limit of this sequence on the completion of the initial space. Therefore, we can obtain the convergence of the arbitrary fundamental sequence with respect to the metric of the completion.

5.6 Conclusions

1. The justification of the determination of mathematical models is based on the passage to the limit.

2. The definition of the limit is not constructive because it uses a priori knowledge of the limit.

3. The proof of the convergence can be based on the Cauchy criterion that uses the fundamentality of the sequences and does not require a priori knowledge of the limit.

4. The Cauchy criterion is applicable for the complete spaces only.

5. The majority of the spaces is non-complete.

6. The classic example of the non-complete spaces is the set of rational numbers.

7. The divergent fundamental sequences of rational numbers determine the irrational numbers by Cantor.

8. The Cantor real numbers are the equivalent classes of the fundamental sequences of rational numbers.

9. The Cantor real numbers satisfy all properties of the axiomatic definition of real numbers.

10. Each irrational number can be approximated by rational numbers.

11. The space of real numbers is complete.

12. Cantor's method of determination of real numbers is the basis of the completion of the arbitrary metric spaces.

13. By the completion theorem, each metric space can be extended to a complete metric space such that all elements of this extension can be approximated by elements of the initial space.

14. By the completion theorem, each fundamental sequence of the arbitrary space is convergent. However, its limits can be elements of the extension of the initial space.

We shall try to use the considered technique that is the sequential method for the justification of mathematical models. However, we consider, at first, different applications of the sequential method.

Notes

[1]Intuitionism is one of the directions of the foundations of mathematics, which appeared in connection with the paradoxes of set theory. The main ideologist of this direction was *Luitzen Egbertus Jan Brouwer*. Intuitionistic mathematics is considered in [68].

[2]Constructive mathematics is considered in [55], [121].

[3]There exist constructive definitions of **natural numbers**. This is based on the **cardinality** that is the general characteristic of the sets, see [27], [55], [77], [99]. Two sets have the same cardinality if and only if there exists one-to-one mapping of one set to another. This is the quantity of the elements for the finite sets. Thus, the first natural number 1 is the of the set $\emptyset$ that includes only the empty set as the element. Let us have a number n that is the cardinality of a set X. Then the next natural number n' is the cardinality of the set X' that is the result of the addition to X of an arbitrary element that does not belong to X. The determined set satisfies, of course, all Peano axioms. The analogical definition was proposed by *Friedrich Ludwig Gottlob Frege* in 1884.

[4]*Giuseppe Peano* proposed the axiomatic definition of the set of natural numbers in 1889.

[5]The first axiomatic definition of a real number was proposed by *David Hilbert* in 1900.

[6]The set of real numbers can be determined as an ordered field with the Dedekind continuity axiom or the maximal ordered Archimedean field; see, for example, [12], [70], [90], [109], [182], [189].

[7]**Abstract algebra** is the branch of algebra, which studies the sets with different operations and the transformations of these sets, see [25], [108], [132], [174], [191] [200].

[8]One also determines the n-order operation. This is the transformation, which maps n arbitrary elements of the given set to an element of the same set. For example, the transformation of the given element of the group to its inverse element is the first order operation. Another example of the first order operation on the set of natural numbers is the transformation of the natural number to the next number (see Example 5.1). We will determine also that the differentiation is the first order operation on the set of infinite differentiable functions and on the set of distributions.

[9]The group theory is one of the most important branches of algebra with many applications; see, for example, [9], [82], [145], [146], [174], [159]. Particularly, this is the basis of the symmetry theory.

[10]The linear (vector) spaces are the algebraic base of the different mathematical spaces (linear topological, normed, Banach, Hilbert, etc. The theory of these spaces is described in [63], [110], [179].

[11]The linear topological spaces are considered, for example, in [28], [81], [93], [144], [153].

[12]The theory of linear normed space, particularly, Banach spaces (complete linear normed space) is an important direction of the functional analysis; see, for example, [71], [81], [91], [95], [142], [147], [203].

[13]The complete unitary space is called **Hilbert**. This is the most important class of mathematical spaces, see [63], [71], [81], [91], [95], [142], [147], [203].

[14]The general concept of the field was proposed by *Évariste Galois* in 1830. The theory of fields is considered in [2], [9], [24], [108].

[15]The ordered sets are considered in [6], [27], [156], [174].

[16]The Archimedean axiom was first formulated by *Eudoxus*.

[17]The real numbers as infinite decimal fraction was determined by *Karl Theodor Wilhelm Weierstrass* in 1866. The Weierstrass real numbers are considered, for example, in [182].

[18]The properties of the Weierstrass real numbers are considered [182].

[19]The principle of nested segments was proved by *Georg Cantor* in 1872.

[20]The Bolzano–Weierstrass theorem was actually first proved by *Bernard Bolzano* in 1817 as a auxiliary result. This theorem proved again by Karl Weierstrass as one of the most important assertion of the mathematical analysis.

[21]The completeness of the set of real number was proved by *Georg Cantor*.

[22]The cuts and the definition of real numbers on the base of the cuts were proposed by *Julius Wilhelm Richard Dedekind* in 1872.

[23]The concept of completeness was proposed by *Maurice René Fréchet* in 1906. The completion theorem was proved by *Felix Hausdorff* in 1914. The proof of the completion theorem is given, for example, in [81], [83], [91], [142].

[24]The completion theorem for uniform spaces is given in [26].

Part III

Sequential objects

Part III

Sequential objects

The necessity of justifying the procedure for constructing mathematical models led us to the study of the problem of convergence of the sequences. The existence of the limit can be proven by the Cauchy criterion. Its direct application is possible only in the case of completeness of the space. We can use the completion procedure for the general situation, which is the basis of the sequential method. The most natural realization of this idea is the definition of real numbers based on the completion of the set of rational numbers.

Before applying this apparatus to construct mathematical models, we will consider other forms of sequential objects. Particularly, the set of real numbers is not the unique extension of the class of rational numbers by the sequential method. The extremely interesting and important class of p-adic numbers is another sequential extension of the set of rational numbers. A qualitatively different application of this method is the optimal control theory, where the sequential method is a means of analyzing extremum problems that do not have any solutions in the natural sense. In the concluding chapter of this part, we consider a sequential interpretation of the set of distributions associated with the extension of the solution of mathematical physics problems. These results are already close to solving the problem of justifying mathematical models.

Chapter 6

p-adic numbers

The justification of mathematical models can be realized by the passage to the limit. One can analyze the convergence of the sequences using the Cauchy criterion, if it is applicable. The completion of the metric spaces, which is the basis of the sequential method, can be applied in the general case. We used this idea before for the determination of the set of real numbers from the given set of rational numbers. However, this method is applicable for the definition of other extensions of rational numbers. Of course, this uses other forms of metric. We determine the sets of p-adic numbers as a result. At first, these objects are determined by means of algebraic constructions. This is based on the comparison of integers modulo. The same result can be obtained with using the completion method for the set of rational numbers with respect to the specific p-adic metric.

6.1   Comparisons of integers modulo

Determine, at first, a very important notion of the number theory and its applications. This is comparison. The comparison of two integers with respect to a natural number m is a special mathematical operation. This allows us to answer the question of whether these two integers have the same remainder when divided by m. Naturally, any integer divided by m gives one of m variants of the possible remainder. There are all natural numbers from 0 to $m-1$. This means that all integers can be divided into m sets. Each number of a concrete set has the same remainder of division by m. As a result, we come to the next concept.

Definition 6.1 *The integers x and y are **comparable of the modulo** m, if these numbers have the same remainder of division by the natural number m.*

The integer numbers x and y are comparable of a modulo m, if their differences are divisible by m. Then there exists an integer k such that the following equality holds

$$x = y + km.$$

Of course, if this equality is true, the considered numbers have the same remainder of division by m. The comparability of the integers x and y of a modulo m is denoted by

$$x \equiv y \pmod{m}.$$

Example 6.1 ***Comparable of the modulo*** 6. The number 7 has the remainder 1 of division by the modulo 6. The numbers 13 and -5 have the same remainders. Therefore, these numbers are comparable of a modulo 6. Thus, we get

$$13 \equiv 7 (\text{mod } 6), \; -5 \equiv 7 (\text{mod } 6).$$

Of course, the differences $13 - 7$ and $(-5) - 7$ are divisible by m. Besides, the following equalities hold

$$13 = 7 + 1 \cdot 6, \; -5 = 7 + (-2) \cdot 6.$$

Thus, the considered representation of the comparable numbers is true. □

Obviously, any integers are comparable of a modulo 1. If the integers x and y are comparable of a modulo m, and the number n is a divisor of m, then the numbers x and y are comparable of the modulo n. The numbers x and y to be comparable of the modulo m with decomposition into prime factors p_1, $p_2, \ldots, p_k$

$$m = \prod_{i=1}^{k} p_i^{\alpha_i},$$

whenever the following comparison hold.

$$x \equiv y \pmod{p_i^{\alpha_i}}, \; i = 1, \ldots, k,$$

where $\alpha_1, \ldots, \alpha_k$ are natural numbers.

Example 6.2 ***Comparable of the modulo*** 6 ***and*** 18. The number 6 has the prime factors 2 and 3. It is obvious that the differences $13 - 7$ and $(-5) - 7$ are divisible by 2 and by 3. Therefore, the following comparisons hold

$$13 \equiv 7 \pmod{2}, \; 5 \equiv 7 \pmod{2};$$

$$13 \equiv 7 \pmod{3}, \; 5 \equiv 7 \pmod{3}.$$

Thus, the comparability of the composite module follows from the comparability of all its factors. Note that the numbers 13 and -5 are comparable not only of the modules 2 and 3, but modulo $9 = 3^2$, since their difference is divided by 9 without a remainder. Then they these numbers are comparable of the modulo $2^1 \cdot 3^2 = 18$. Indeed, the difference between 13 and -5 is just 18. Thus, the comparability of integers by powers of some prime numbers does have their comparability of the product of these degrees. □

The comparability of a modulo m is the relation on the set of integer numbers.

Theorem 6.1 *The comparability of a modulo m is the equivalence.*

Proof. For any natural number m the comparison

$$x \equiv x \pmod{m}$$

is true because the difference $x - x$ is equal to zero. Therefore, the comparability relation is reflexive. Then from the comparability of numbers x and y it follows the divisibility of its difference by m. Hence, the difference $y - x$ is divisible by m too. Therefore, the comparability is the symmetric relation. Finally, suppose the conditions

$$x \equiv y \pmod{m}, \quad y \equiv z \pmod{m}.$$

Using the formula of the representation of comparable numbers, prove the existence of integers k and l such that

$$x = y + km, \quad y = z + lm.$$

Then the following equality holds

$$x = z + (k + l)m.$$

Thus, the integers x and z are comparable of the modulo m. Then the considered relation is transitive. This completes the proof of the theorem. □

Definition 6.2 *The set of all integers that are comparable with number a of the modulo m is called the **residue class** a of the modulo m. Each number of this class is called the **residue** of the modulo m with respect to any number of this class.*

The considered residue class is denoted by $[a]_m$. The set of all residue classes of the modulo m is denoted by $\mathbb{Z}_m$. Obviously, the residue class is the equivalence class. Therefore, the set $\mathbb{Z}_m$ is the factor-set of the set of integers with respect to the relation of the comparison of the modulo m. The quantity of these residue classes is equal to the quantity of the possible residues of the division by m. These residues are the numbers $0, 1, \ldots, m - 1$. Thus, the set $\mathbb{Z}_m$ consists of m elements.

The numbers x and y are comparable of the modulo m whenever the equality of the residue classes $[x]_m = [y]_m$ is true.

TABLE 6.1:   Addition on
the set $\mathbb{Z}_6$

	[0]	[1]	[2]	[3]	[4]	[5]
[0]	[0]	[1]	[2]	[3]	[4]	[5]
[1]	[1]	[2]	[3]	[4]	[5]	[0]
[2]	[2]	[3]	[4]	[5]	[0]	[1]
[3]	[3]	[4]	[5]	[0]	[1]	[2]
[4]	[4]	[5]	[0]	[1]	[2]	[3]
[5]	[5]	[0]	[1]	[2]	[3]	[4]

Remark 6.1 The elements of the arbitrary set are equivalent with respect to each equivalence relation if and only if the correspondent equivalence classes are equal.

Definition 6.3 *The **complete residue system** of the modulo m is a system of m pairwise incomparable with respect to these modulo integers.*

One chooses usually the least non-negative residues $0, 1, \ldots, m-1$ as the complete residue system. The operations of addition and multiplication on the set $\mathbb{Z}_m$ are determined by the equalities[1] (see Figure 6.1)

$$[a]_m + [b]_m = [a+b]_m,$$

$$[a]_m \cdot [b]_m = [ab]_m.$$

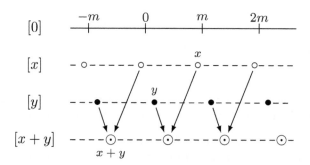

FIGURE 6.1: Addition of the residue classes.

Remark 6.2 The residue system permits us to realize the arithmetic operations on the finite set.

Example 6.3 *Operations on the set $\mathbb{Z}_6$.* Addition and multiplication on the set $\mathbb{Z}_6$ is given in Table 6.1 and Table 6.2. The index denoting the modulo is not given here for brevity. $\square$

TABLE 6.2: Multiplication on the set $\mathbb{Z}_6$

	[0]	[1]	[2]	[3]	[4]	[5]
[0]	[0]	[0]	[0]	[0]	[0]	[0]
[1]	[0]	[1]	[2]	[3]	[4]	[5]
[2]	[0]	[2]	[4]	[0]	[2]	[4]
[3]	[0]	[3]	[0]	[3]	[0]	[3]
[4]	[0]	[4]	[2]	[0]	[4]	[2]
[5]	[0]	[5]	[1]	[2]	[3]	[4]

TABLE 6.3: Multiplication on the set $\mathbb{N}_6$

	0	1	2	3	4	5
0	0	0	0	0	0	0
1	0	1	2	3	4	5
2	0	2	4	0	2	4
3	0	3	0	3	0	3
4	0	4	2	0	4	2
5	0	5	1	2	3	4

Definition 6.4 *The set X with the operations $+$, and $\cdot$ is called the **ring**[2], if this set with operation $+$ is an abelian group, and the following equalities hold*

$$x \cdot (y \cdot z) = (x \cdot y) \cdot z, \quad x \cdot (y + z) = (x \cdot y) + (x \cdot z), \quad (y + z) \cdot x = (y \cdot x) + (z \cdot x)$$

for all $x, y, z \in X$.

The set $\mathbb{Z}_6$ with considered operations is the ring.

Remark 6.3 This is the factor-set of the set of integer numbers with respect to the comparability of the modulo m. The factor-set is the partial case of the **factor-object**[3] that can be determined on the basis of determining of the natural structure of the initial set on the factor-set. The factor-group, the factor-topology, etc. are other examples of the factor-objects.

Consider an operator A that maps the ring $\mathbb{Z}_m$ to a complete residue system, for example, $\mathbb{N}_m = \{0, 1, ..., m-1\}$. This operator is invertible. Determine the arithmetic operations on the set $\mathbb{N}_m$

$$x + y = [A^{-1}x]_m + [A^{-1}x]_m,$$

$$x \cdot y = [A^{-1}x]_m \cdot [A^{-1}x]_m.$$

Particularly, the multiplication on the set $\mathbb{N}_6$ is determined in the Table 6.3. The operations on the sets $\mathbb{Z}_m$ and $\mathbb{N}_m$ are the same properties. Therefore, the rings $\mathbb{Z}_m$ and $\mathbb{N}_m$ are **isomorphic**, and the map A between these rings is an **isomorphism**.

TABLE 6.4:
Multiplication on the set $\mathbb{Z}_5$

	[0]	[1]	[2]	[3]	[4]
[0]	[0]	[0]	[0]	[0]	[0]
[1]	[0]	[1]	[2]	[3]	[4]
[2]	[0]	[2]	[4]	[1]	[3]
[3]	[0]	[3]	[1]	[4]	[2]
[4]	[0]	[4]	[3]	[2]	[1]

Remark 6.4 The isomorphism is one of the general mathematical concepts[4]. We considered isomorphisms also for other object classes. For example, we used the isometry that is the isomorphism of the metric spaces. The isomorphic objects (the rings, the metric spaces, etc.) have the same properties that can be determined by the subject area (the ring theory now, the metric space theory before). Thus, the isomorphic objects do not differ in a concrete particular theory. Many mathematical results are established up to corresponding isomorphisms.

The considered ring $\mathbb{Z}_6$ has some very unusual properties. For example, we have the equalities

$$[1]_6 \cdot [3]_6 = [3]_6, \quad [3]_6 \cdot [3]_6 = [3]_6, \quad [5]_6 \cdot [3]_6 = [3]_6.$$

Therefore, the class $[3]_6$ is the result of multiplication by $[3]_6$ of three classes. There are $[1]_6$, $[3]_6$, and $[5]_6$. Then we have the question. What do we have after division of the class $[3]_6$ by $[3]_6$? Thus, the division by $[3]_6$ on the ring $\mathbb{Z}_6$ is not applicable, because we can obtain three objects as the result. Note (see Caption 4), the ring can be the field whenever we cannot do divide by each non-zero element. The class $[0]_6$ is zero on the considered ring. Therefore this is not the field because we cannot divide by a non-zero elements of the ring $\mathbb{Z}_6$.

We have the question, can the ring $\mathbb{Z}_m$ be a field under certain conditions?

Example 6.4 *Ring* $\mathbb{Z}_5$. The results of multiplication on the ring $\mathbb{Z}_5$ are given in Table 6.4. The index denoting the modulo is not given again. Note that the division is determined now for all non-zero elements of the ring. For example, after division of $[1]_5$ by $[2]_5$ we have the class $[3]_5$; dividing the class $[3]_5$ by $[4]_5$, we get $[2]_5$. It is not very difficult to prove that the ring $\mathbb{Z}_5$ satisfies the definition of the field. □

The difference between the properties of the rings $\mathbb{Z}_5$ and $\mathbb{Z}_6$ is the corollary of the properties of its modulo. Particularly, the number 5 is prime; and the number 6 is composite. The ring $\mathbb{Z}_m$ is the field whenever the number m is prime. Therefore, the residue classes with prime modulo are most interesting. These classes are the basis of the definition of p-adic numbers.

6.2 Integer *p*-adic numbers

Determine *p*-adic numbers by two steps. At first, we consider the integer *p*-adic numbers. Then we determine general *p*-adic numbers using the technique of definition of rational numbers on the basis of integers.

Definition 6.5 *The **integer p-adic number** is the infinite sequence* $x = \{x_1, ..., x_k, ...\}$ *of residues* x_k *of the* p^k, *where the number p is prime such that*

$$x_k \equiv x_{k+1} \ (mod \ p^k). \tag{6.1}$$

Obviously, each integer x can be interpreted as the *p*-adic number $\{x, ..., x, ...\}$ that is the stationary sequence. Indeed, all elements of this sequence are equal. Therefore, these are comparable of each modulo.

Note, that *p*-adic numbers are represented by the sequences not unequivocally. The sequence $\{x_k\}$ and $\{y_k\}$ determine the same *p*-adic number, if

$$x_k \equiv y_k \ (\text{mod } p^k).$$

Try to determine the easiest representation of a *p*-adic number.

Let the sequence $\{x_k\}$ determine a given *p*-adic number. Try to find its easy enough representation. Let us consider the first element x_1 of this sequence. There exists a unique comparable of the modulo p the number a_1 such that

$$0 \leq a_1 < p.$$

This is a digit of the *p*-ary number system. Choose it as the element of the easy sequence $\{y_k\}$ that determines the same considered *p*-adic number. Determine $y_1 = a_1$; this is an easy enough integer that is comparable with x_1.

The second element of the sequence $\{x_k\}$ satisfies the condition (6.1) for $k = 1$. Therefore, the numbers x_1 and x_2 are comparable of the modulo p. Then these integers have the same final digit of the *p*-ary number system. Its difference has in this system a zero low order that is the sign of the divisibility by p. Then there is the low order of the number a_1. Denote by a_2 the second digit on the right of the *p*-ary form of the number x_2. Therefore, each integer y_2 that is comparable of modulo p^2 has the two last *p*-ary digits equal to a_2a_1. The easiest of these numbers is $a_1 + a_2p$. This two-digit number we choose as y_2.

The third element of the sequence $\{x_k\}$ satisfies the comparison (6.1) for $k = 2$. Therefore, the numbers x_2 and x_3 are comparable of the modulo p^2. These numbers have the same two last digits of the *p*-ary number system; they end with a_2a_1. Denote by a_3 the third digit right of *p*-ary number x_3. Then the number y_2 that is comparable of the modulo p^3 has the last three *p*-ary digits $a_3a_2a_1$. The easiest of these numbers is three-digit and equals $a_1 + a_2p + a_3p^2$. We choose it as y_3.

Thus, we have the concrete sequence $\{y_k\}$ that represents the considered p-adic number. It is easy enough, because its elements satisfy the inequality

$$0 \le y_k < p^k.$$

Each element is determined by adding a digit at the left of the previous element. This representation of the p-adic number is called **canonic**. The number y_k in the k-ary system is the p-digit number

$$y_k = a_k a_{k-1}...a_2 a_1,$$

where a_i are digits of this system, $i = 1, ..., k$. Therefore, each p-adic number can be determined uniquely by the sequence

$$x = \{a_1, a_2 a_1, a_3 a_2 a_1, ..., a_k a_{k-1}...a_2 a_1, ...\}.$$

This object can by represented formally as infinite in the left sequence of digits of the p-ary system

$$x = ...a_k a_{k-1}...a_2 a_1.$$

In reality, this is the representation of the p-adic number as the series

$$x = \sum_{i=1}^{\infty} a_i p^i.$$

Remark 6.5 The last formula is an analogue of the function representation as the *Fourier series* with degrees of the prime p as the basis and the digits a_i as the Fourier coefficients.

Example 6.5 *Natural numbers.* Suppose $p = 5$. Determine the 5-adic representation of the number 8. This is the sum $1 \cdot 5 + 3$. Thus, the number 8 can be written as 13 in the 5-ary system. Prove the 5-adic representation

$$8 = ...00013.$$

Indeed, a_i is the i-th digit of this representation. Then we have

$$a_1 = 3, \ a_2 = 1, \ a_3 = a_4 = ... = 0.$$

Using the considered technique, determine

$$x_1 = a_1 = 3, \ x_2 = a_2 a_1 = 13, \ x_3 = a_3 a_2 a_1 = 013, ...$$

The number $x_1 = 3$ and $x_2 = 13$ are comparable of the modulo 5, because the absolute value of its difference is the 5-ary number 10 that is 5. The numbers x_2 and x_3 are actually equal; hence, these numbers are comparable of the modulo 25. Prove analogically that the condition (6.1) is true for all k. Each natural number can be represented as the infinite in the left sequence of p-ary digits with finite quantity of the non-zero digit. The number 0 is represented by the sequence, where all elements are zero. $\square$

TABLE 6.5: Representation of usual integer number in integer rings of p-adic numbers

$\mathbb{Z}$	$\mathbb{Z}_2$	$\mathbb{Z}_3$	$\mathbb{Z}_5$	$\mathbb{Z}_7$
4	...00100	...00011	...00004	...00004
10	...01010	...00101	...00020	...00013
-3	...11101	...22220	...44442	...66664

Determine the addition and the multiplication on the set of integer p-adic numbers by the equalities

$$x + y = \{x_1 + y_1, ..., x_k + y_k, ...\},$$

$$x \cdot y = \{x_1 y_1, ..., x_k y_k, ...\},$$

where $x = \{x_1, ..., x_k, ...\}$, $y = \{y_1, ..., y_k, ...\}$. The set of integer p-adic numbers with these operations is the ring that is denoted by $\mathbb{Z}_p$. Besides, the set of integers $\mathbb{Z}$ is the subring of the ring $\mathbb{Z}_p$.

Remark 6.6 The usual integer numbers are called often the *integer rational numbers*.

Using the representation of p-adic numbers by the sequence p-ary digits, we can determine the natural operation of the addition, multiplication and difference. The realization of these operations is bit-by-bit with respect to the corresponding representations. Determine the representation of negative numbers by means of this technique.

Example 6.6 *Negative numbers*. Negative numbers can be determined as a result of the difference from zero of a corresponding positive number. Determine, for example, the 7-adic representation of the number -5. We have the equality $-5 = 0 - 5$. The number 0 has the 7-adic infinite representation ...0000; and the number 5 is represented by ...0005. After the difference of these values in the 7-ary system, we have ...6662. Indeed, adding bit-by-bit of this object with ...0005 that is the number 5, determine the sequence of zero that is the representation of the number 0. We can prove that each negative number has the p-adic representation with the infinite set of non-zero digits. □

The representations of some integers in the different ring $\mathbb{Z}_p$ are determined in Table 6.5.

Now we can define the general p-adic numbers that are not necessarily integers by means of the well-known method of definition of rational numbers on the basis of integers.

6.3 General p-adic numbers

The extension of the set of usual integer numbers to the set of rational numbers is related to the problem of the divisibility of integers. This is the problem of the impossibility of the inversion for the integer numbers with respect to multiplication. Consider the invertibility of integer p-adic numbers.

Example 6.7 *Invertibility of integer p-adic numbers that are divisible by p.* Consider a p-adic number that is divisible by p, for example, the number p. It has the form ...00010. Suppose there exists a p-adic number with representation $x = ...a_k a_{k-1}...a_2 a_1$ that is inverse to p. Then we get the equality $xp = 1$. The last number has the representation ...00001 here. Thus, it is necessary to find the p-ary digits a_1, a_2,... such that the product of ...$a_k a_{k-1}...a_2 a_1$ by ...00010 is ...00001. Note, that the result of multiplication of its low digits $a_1 \cdot 0$ gives the low digit of the representation of the number 1 that is 1. Of course, this is impossible. Therefore, the number p is not invertible with respect to multiplication in the ring $\mathbb{Z}_p$. One can prove that each number that is divisible by p is not invertible in this ring. $\square$

Example 6.8 *Invertibility of integer p-adic numbers that are not divisible by p.* Consider a concrete example. Try to invert the usual number 6 as 5-adic in the ring $\mathbb{Z}_5$. The number 6 is 11 in the 5-ary system. Therefore, it has the 5-adic representation ...000011. Suppose a number $x = ...a_k a_{k-1}...a_2 a_1$ is inverse to ...000011. Realize the multiplication of these numbers (see Table 6.6). The first line here (the result of the multiplication of x by the final digit 1 of the representation of the number 6) is ...$a_4 a_3 a_2 a_1$. The second line (the result of the multiplication of the penultimate digit 1 of the representation of the number 6) is the same sequence that is shifted one digit to the left. All of the next lines consist of zero only, because two digits of 6 only are characterized by non-zero digits. By Table 6.5 the digit a_1 of the final 5-ary form of the number 6 is 1. Then the penultimate digit of the addition of the one-digit numbers a_2 and a_1 that is 1 gives 0. This is possible for $a_2 = 4$ only. Then, as a result of the addition of the digits from the second from the end of the terms gives the 5-ary number 10. Thus, the penultimate digit of the product is, in reality, 0; and the unit goes to the next (third from the right) digit. We add here the digits a_3 and a_2 that are equal to 4 and the unit from the previous position. Then we get 0 that is the third from the right digit of the representation of the number 1. Therefore, the result of the addition is the 5-ary number 10; and $a_3 = 0$. Now we add the numbers a_4, a_3 that is equal 0, and the unit from the previous position. The fourth digit from the left of the representation of the number 1 in 0. Then we have $a_4 = 4$. Repeating this process, we conclude that all the next digits of the 5-adic representation of the number x are 0 and 4 only. Therefore, we find $x = ...4040401$. Thus, the number 6 is invertible in the ring $\mathbb{Z}_5$. We can prove that any non-divisible by p number is invertible in $\mathbb{Z}_p$. $\square$

TABLE 6.6:
Representation of usual
integer number in integer
rings of *p*-adic numbers

numbers	5-adic representation
x	$...a_4a_3a_2a_1$
6	$...0\ 0\ 1\ 1$
+	$...a_4a_3a_2a_1$ $...a_3a_2a_1$ $0\ 0$
1	$...0\ 0\ 0\ 1$

Remark 6.7 Note the serious difference between the rings of the usual and *p*-adic integer numbers. It is possible to divide by 1 only for the set of the usual integer numbers. However, it is possible to divide by each non-divisible by *p* number for the ring of integer *p*-adic numbers *p*.

Remember the definition of the set of rational numbers on the basis of the given set of integers for the extension of the ring of -adic numbers. Consider the equation

$$b \cdot x = a \qquad (6.2)$$

that is called the ***multiplication equation***, where the parameters a and b are integer; besides b is not equal to zero. This equation does not have an integer solution x for all integer parameters. We could guarantee its solvability by the extension of the set of integer numbers. This set can be extended by adding all desired solutions of the multiplication equation. The set of rational numbers is the result of this idea. The pair (a, b) here is a rational number that is the solution of the equation (6.2). Note, that a non-unique pair of integers can determine each rational number. Give the stricter definition.

Definition 6.6 *The set of **rational numbers** is the set $\mathbb{Q}$ of the equivalence classes of pairs (a, b) of integers, where b is non-zero. The pairs (a, b) and (a', b') are equivalent here whenever $ab' = a'b$.*

For example, the pairs $(1, 2)$ and $(2, 4)$ are equivalent. These pairs determine the same rational number that is denoted by $1/2$. The rational number that is determined by the pair (a, b) is denoted by the fraction a/b. The integer numbers can be interpreted as the rational numbers $(a, 1)$ that are the fractions $a/1$.

Determine the addition and the multiplication on the set of rational numbers using the analogical operations of integers by the equalities

$$a_1/b_1 + a_2/b_2 = (a_1b_2 + a_2b_1)/(b_1b_2),$$

$$a_1/b_1 \cdot a_2/b_2 = (a_1a_2)/(b_1b_2).$$

Remark 6.8 Analogically, we can determine the ***integer numbers*** as the equivalence classes of pairs of natural numbers.

The set $\mathbb{Q}$ with these operations is the field. Each non-zero rational number a/b is invertible here with respect to multiplication. The corresponding inverse element is the fraction b/a. Note, that then the multiplication equation (6.2) has the unique solution for all rational values a and b if $b \neq 0$. This is the result of the division of the number a by b that is the multiplication of a by the inverse number to b.

This technique can be extended from the concrete ring $\mathbb{Z}$ to the large enough class of fields. Consider a commutative ring with unit element with respect to the multiplication such that the product of each two non-zero elements is not equal to zero. This is called **entire ring**. The integer ring $\mathbb{Z}$ and integer p-adic rings $\mathbb{Z}_p$ are entire.

Remark 6.9 The considered ring $\mathbb{Z}_6$ is commutative. It has also the unit [1]. However, it is not entire. Particularly, the product of non-zero elements [2] and [3] (not only their product, see Table 6.2) is equal to [0] that is the zero element of this ring (see Table 6.1). By the way, the non-zero elements of the ring with zero product are called the **divisors of zero**.

Definition 6.7 *The **field of quotients** of the entire ring is the set of equivalence class of pairs (a, b) of this set with a non-zero second element, where the pairs (a, b) and (a', b') are equivalent whenever the following equality holds $a \cdot b' = a' \cdot b$. The operation of addition, difference, multiplication, and division by a non-zero element is determined by standard formulas.*

Definition 6.6 is the partial case of this definition with the entire ring of integer numbers.

Now we can determine the general p-adic numbers[5].

Definition 6.8 *p-adic numbers* are the elements of the field of quotient $\mathbb{Q}_p$ of the ring $\mathbb{Z}_p$ of integer p-adic numbers.*

Thus, it is possible to realize the standard arithmetic operations on the p-adic numbers. We do not leave the set $\mathbb{Q}_p$ after addition, difference, multiplication, and division by a non-zero element. The extension of the ring of integer p-adic numbers is realized by means of the division by the numbers that are divisible by p. Note the usual negative degrees of the base of number system; for example, 10^{-1}, can be represented as the corresponding fractions, for example, 0,1. This result is true for the p-adic numbers too.

Example 6.9 *Inversion of the number p in $\mathbb{Q}_p$.* The number p has the p-adic representation ...00010. Using decimal value 0,1 (zero integer and one-ten) write formally the inverse to p number p^{-1} as ...000,1. Try to multiply these numbers (see Table 6.7). The result is ...001 that is the p-adic representation of the number 1. Therefore, the value ...000,1 can be actually interpreted as the p-adic representation of the number p^{-1}. $\square$

Each usual non-negative rational number x can be written in the p-ary number system by the formula

$$x = \sum_{i=-s}^{k} a_i p^{i-1},$$

TABLE 6.7: *p*-adic multiplication of the numbers p and p^{-1}

numbers	5-adic representation
p^{-1}	...0000,1
p	...0010
$+$	...0000,0 ...0001 ...000
1	...0001

or shorter

$$x = a_k a_{k-1}...a_2 a_1, a_0 a_{-1}...a_{-s}$$

where a_i are p-ary digits, and s is a natural number. The negative numbers have the analogical form with the minus sign. The p-adic numbers, i.e. the elements of the fields $\mathbb{Q}_p$ can be represented by the formula

$$x = \sum_{i=-s}^{\infty} a_i p^{i-1},$$

or

$$x = ...a_k a_{k-1}...a_2 a_1, a_0 a_{-1}...a_{-s}$$

Of course, each rational number is p-adic. The relations between different numerical classes is given by Figure 6.2. The arithmetic operations between these numerical representations can be carried out naturally. Note, that each p-adic number of the field $\mathbb{Q}_p$ can be transformed to the product xp^n, where x is not divisible by p, and n is a usual integer number.

Remark 6.10 It is curious that in the interpretation of real numbers by Weierstrass the number is represented as an infinite decimal fraction. It is characterized by an infinite number of digits on the right, separated by the symbol ",", to distinguish the zero digit. Now we have the infinite number of digits on the left.

Remark 6.11 The representation of the p-adic numbers as a series with a finite set of negative indexes is an analogue of the representation of the function by the Laurent series[6].

6.4 *p*-adic metrics

Return to the consideration of the set of rational numbers. Determine a special functional here. Each rational number x can be transformed to the value

$$x = a/bp^n,$$

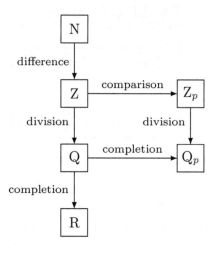

FIGURE 6.2: Relations between numerical classes.

TABLE 6.8:   2-adic and
3-adic norms of rational numbers

x	p	n	$\|x\|_p$	p	n	$\|x\|_p$
2	2	1	1/2	3	0	1
1/18	2	-1	1	3	-2	9
-54	2	1	1/2	3	3	1/27

where a/b is the irreducible fraction, besides a and b are integer numbers that are not divisible by the prime number p, and n is integer.

Definition 6.9 *The **p-adic norm** of the non-zero rational number* $x = a/bp^n$ *is the number*

$$\|x\|_p = p^{-n};$$

the p-adic norm of zero is equal to zero.

Remark 6.12 We determined before the norm on the linear spaces (see Caption 3). Now we do not interpret the set of rational numbers as a linear space. Therefore, the p-adic norm has a few other senses.

The p-adic norm characterizes the degree of divisibility of the given rational number x by the prime number p. This norm decreases by increase of the exponent n of the degree of divisibility. The 2-adic and 3-adic norms of rational numbers are given in Table 6.8.

Note the multiplicativity of the p-adic norm

$$\|xy\|_p = \|x\|_p\|y\|_p.$$

Determine the functional ρ_p on the set of pairs of rational numbers by the equality

$$\rho_p(x, y) = \|x - y\|_p.$$

Analyze the properties of this functional. It has non-negative values only. The p-adic norm can be equal to zero for the zero number only. Therefore, the equality $\rho_p(x, y) = 0$ is true for $x = y$ only. The equality $\rho_p(x, y) = \rho_p(y, x)$ is true, because the p-adic norm does not depend on the sign of the rational number. There are the properties of the metric. This is the p-**adic metric**. However, the metric satisfies the triangle inequality extra. Prove the following result.

Lemma 6.1 *The functional ρ_p satisfies the inequality*

$$\rho_p(x, y) \leq \max \left\{ \rho_p(x, z), \rho_p(y, z) \right\} \ \forall x, y, z \in \mathbb{Q}. \tag{6.3}$$

Proof. Determine $\varphi = xz$, $\psi = zy$. Then $\varphi + \psi = xy$. The condition (6.3) is true whenever the following inequality equality holds

$$\|\varphi + \psi\|_p \leq \max \left\{ \|\varphi\|_p, \|\psi\|_p \right\}. \tag{6.4}$$

Assume for definiteness that the p-adic norm of ψ is not greater than the p-adic norm of φ. Therefore, we have the equalities

$$\varphi = a/bp^n, \quad \psi = c/dp^{n+k},$$

where the integers a, b, c, d are not divisible by p, n is integer, and k is a non-negative integer number. Find the sum

$$\varphi + \psi = \left(adp^k + cd \right)/bd\, p^n.$$

The product bd is not divisible by p, but the sum $adp^k + cd$ can be divisible by p. Thus, the exponent of p of the sum $\varphi + \psi$ is not less than n. Therefore, the p-adic norm of $\varphi + \psi$ is not greater than $-n$. However, the norm of φ is equal to $-n$, and the norm of ψ is not greater than $-n$. Hence, the inequality (6.4) is true, and the condition (6.3) too. This completes the proof of the lemma. $\square$

Obviously, the triangle inequality follows from the condition (6.3). Thus, the set of rational numbers is the metric space with p-adic metric ρ_p.

Remark 6.13 The inequality (6.3) with previous properties of the metric corresponds to the notion of **ultrametric**, and a set with an ultrametric is called the **ultrametric space**[7]. Among all metric spaces, ultrametric spaces are distinguished by very unusual properties. In particular, all triangles of the ultrametric space are isosceles; that is, at least two of their sides coincide (see Figure 6.3). The Archimedes axiom is false in the ultrametric space; i.e. there is no guaranteed possibility, repeatedly postponing a smaller segment, to obtain a segment of a greater length. Finally, what is extremely surprising is that every point of the ball there is its center.

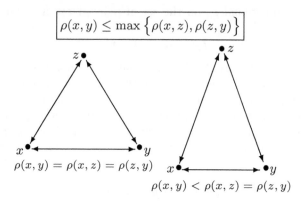

FIGURE 6.3: The triangles of the ultrametric space are isosceles.

We can determine many different metrics on a set. However, several metrics on the same set can be comparable.

Definition 6.10 *The metrics ρ_1 and ρ_2 on a set X are called **equivalent**, if there exist positive constants c_1 and c_2, such that*

$$c_1 \rho_1(x,y) \leq \rho_2(x,y) \leq c_2 \rho_1(x,y) \ \forall x,y \in X.$$

For example, the following metrics of the n-dimensional Euclid space $\mathbb{R}^n$

$$\rho_\infty(x,y) = \max_{i=1,\dots,n} |x_i - y_i|; \ \ \rho_m(x,y) = \left[\sum_{i=1}^m |x_i - y_i|^m \right]^{1/m}, \ m \geq 1$$

are equivalent. By Definition 6.10, if a sequence converges or diverges with respect to a metric, the analogical result is true with respect to each equivalent metric.

Remark 6.14 The sets with equivalent metrics are isometric spaces; i.e. these spaces do not differ in their metric properties.

The question arises whether the natural and p-adic metrics on the set of rational numbers will be equivalent. Consider the properties of easy enough sequences of rational numbers.

Example 6.10 *Stationary sequence.* Consider a stationary sequence $\{x_k\}$, where $x_k = x$ for all k. This sequence tends to the number x with respect to the natural metric of the set of rational numbers. Check this property for the p-adic metric. Determine

$$\rho_p(x_k, 0) = \|x_k - x\|_p = \|0\|_p = 0.$$

Thus, the stationary sequence tends to its natural limit with respect to the p-adic metric too. $\square$

Example 6.11 *Oscillation sequence*. Consider an oscillation sequence, for example the sequence $\{x_k\}$, determined by the equality $x_k = (-1)^k$. This is a divergent sequence with respect to the usual metric. Suppose a rational number x is its limit with respect to the p-adic metric. The metric $\rho_p(x_k, x)$ can have the value $\|-1-x\|_p$ or $\|1-x\|_p$. The sequence $\{\rho_p(x_k, x)\}$ does not tend to zero, because the numbers $-1-x$ and $1-x$ cannot equal to zero simultaneously. Thus, the considered sequence $\{x_k\}$ diverges with respect to the metric ρ_p. □

Example 6.12 *Sequence $\{p^{-k}\}$*. Consider the sequence $\{x_k\}$, determined by the equalities $x_k = p^{-k}$. This sequence has the zero limit with respect to the usual metric of rational numbers. Find the value

$$\rho_p(x_k, 0) = \|x_k - 0\|_p = \|p^{-k}\|_p = p^k.$$

This number does not tend to zero as $k \to \infty$. Thus, the number 0 is not its limit. However, maybe this sequence has another limit $x = a/b\, p^n$. Find

$$\rho_p(x_k, x) = \|p^{-k} - a/b\, p^n\|_p.$$

We have the inequality $n > (-k)$ for large enough value k. Then we obtain

$$p^{-k} - a/bp^n = (b - ap^n + k)/bp^{-k}.$$

The number b is not divisible by p here. The value $b - ap^n + k$ is not divisible by p because this is the difference between two integers, the second of them is divisible by p and the first of them is not divisible. Now we find

$$\rho_p(x_k, x) = p^k$$

for large enough k. Therefore, sequence $\{x_k\}$ does not have any limits as $k \to \infty$. □

Example 6.13 *Sequence $\{p^k\}$*. Consider the sequence $\{x_k\}$, determined by the equalities $x_k = p^k$. This is the divergent sequence with respect to the usual metric. Find the norm

$$\|x_k\|_p = \|p_k\|_p = p^{-k}.$$

This value is infinitely decreased after the increasing of k. We could suppose that the sequence $\{x_k\}$ tends to zero. Determine

$$\|x_k - 0\|_p = \|p^k\|_p = p^{-k}.$$

Passing to the limit as $k \to \infty$, determine the convergence of the sequence $\{x_k\}$ to zero. □

TABLE 6.9:   Convergence of sequences
$\{x_k\}$ with respect to the different metrics.

sequence $\{x_k\}$	usual metric	p-adic metric
$x_k = x$	converges to x	converges to x
$x_k = (-1)^k$	diverges	diverges
$x_k = p^{-k}$	converges to 0	diverges
$x_k = p^k$	diverges	converges to 0

TABLE 6.10:   Analysis of the
convergence of sequences with respect to the
different p-adic metric

sequence $\{x_k\}$	metric ρ_p	metric ρ_q
$x_k = p^k$	converges to 0	diverges
$x_k = q^k$	diverges	converges to 0

The analysis of the convergence problem for the easiest sequences with respect to the standard and p-adic metrics is given in Table 6.4.

The existence of the sequence that converges with respect to first metric and diverges with respect to the second one and the sequence with inverse properties prove non-equivalence of the usual and p-adic metric. However, we have the question about the equivalence of the p-adic metrics with different prime numbers.

Example 6.14 *Power sequences with different base.* Determine the sequences $\{x_k\}$ and $\{y_k\}$ by the equalities

$$x_k = p^k, \quad y_k = q^k,$$

where p and q are different prime numbers. We know (see Example 6.13) that the sequence $\{x_k\}$ tends to zero with respect to the metric ρ_p, and $\{y_k\}$ has the analogical property with respect to the metric ρ_q. Check the convergence of these sequences with using other metrics. Find the value

$$\rho_q(x_k, 0) = \|x_k - 0\|_q = \|p^k\|_q = 1,$$

because the elements of the sequence $\{x_k\}$ do not have any multipliers that are divisible by positive or negative degree of q. One can prove that this sequence does not have any limits with respect to the metric ρ_q. Therefore, the sequence $\{x_k\}$ is convergent with respect to the metric ρ_p and divergent with respect to the metric ρ_q. It is obvious that the sequence $\{y_k\}$ has the inverse properties.
□

Thus, the convergence and the divergence of sequences of rational numbers depend on the concrete p-adic metric (see Table 6.4). Therefore, the different p-adic metrics are not equivalent.

Remark 6.15 By the *Ostrowski theorem*[8], each norm on the set of rational numbers is equal to the usual norm or to one of p-adic norms. The definition of the equivalence of the norms is analogical to this property for the metric here (see Definition 6.10). Note also the *adel formula* that give the relations between all non-equivalent norms for the rational numbers

$$\prod_p \|x\|_p = 1,$$

where the product is realized by all prime numbers p and the value $p = \infty$ for the usual metric.

Now we can give the sequential definition of the p-adic numbers.

6.5 Sequential definition of p-adic numbers

We determine the set of real numbers as the completion of the set of rational numbers with the usual metric. Try to use the analogical method for this set with p-adic metrics. At first, we have the question: can we go beyond the set of rational numbers on the basis of the same procedure with the p-adic metric?

Example 6.15 *Fundamental sequence with respect to a p-adic metric*
Determine the sequence $\{x_k\}$ of rational numbers by the equalities

$$x_k = \sum_{i=-s}^{k} a_i p^{i-1}, \ k = 1, 2, ...,$$

where s is a natural number, and a_i are p-ary digits. Find the difference

$$x_{k+r} - x_k = \sum_{i=k+1}^{k+r} a_i p^{i-1} = \sum_{i=1}^{r} a_{k+i} p^{i-1} p^k.$$

The value before p^k at the right-hand side of this equality is the sum of the degree p^{i-1} with corresponding coefficients. All coefficients except the first are divisible by p, if a_{k+i} is not zero. Then the whole sum is not divisible by p; and the norm of this difference is equal to p^{-k}. If a_{k+t} is the first non-zero value of the digits a_{k+i} at the right-hand side of the last equality, then we get

$$x_{k+r} - x_k = \sum_{i=t}^{r} a_{k+i} p^{i-1} p^{k+t},$$

where the sum before p^{k+t} is not divisible by p. Indeed, its first term only is not divisible by p. Then we get

$$\|x_{k+r} - x_k\|_p = p^{-(k+t)}.$$

Passing to the limit as $k \to \infty$, prove the fundamentality of the sequence $\{x_k\}$ on the set of rational numbers with metric ρ_p. However, we obtain the infinite series after passage to the limit. This is not a rational number. Then this fundamental sequence of rational numbers diverges with respect to the p-adic metric. Therefore, the set $\mathbb{Q}$ with metric ρ_p is non-complete space. $\square$

Now we can use the completion technique for the extension of the set of rational numbers.

Definition 6.11 *The **set of p-adic numbers** is the completion of the set of rational numbers with respect to the p-adic metric.*

Remark 6.16 p-adic numbers have important applications in the theory of Diophantine equations[9]. It may seem surprising, but such seemingly abstract objects as numbers have a physical sense[10].

We determined at first the set of real numbers in the previous section without using the sequential method. Then we constructed the completion of the set of rational numbers using its usual metric. Finally, we proved that this set and the previously determined set of real numbers satisfy the same properties. Thus, we have the same mathematical object with two different interpretations. Now we give the same analysis. We will determine the properties of the set from Definition 6.11 and we will verify that these properties fully coincide with the properties of the set of p-adic numbers given by Definition 6.8.

Using the completion technique, determine the set F_p of all fundamental sequences of rational numbers with p-adic metric. The sequences $\{x_k\}$ and $\{y_k\}$ are equivalent with respect to the relation φ_p here, if $\rho_p(x_k, y_k) \to 0$ as $k \to \infty$.

Example 6.16 *Equivalent fundamental sequences.* Consider the sequence $\{x_k\}$ from Example 6.15. Determine the sequences of rational numbers $\{y_k\}$ and $\{z_k\}$ by the equalities

$$y_k = \sum_{i=s}^{k+1} a_i p^{i-1}, \quad z_k = \sum_{i=s}^{2k} a_i p^{i-1}, \quad k = 1, 2, \dots.$$

Find the values

$$\rho_p(x_k, y_k) = \left\| a_k p^k \right\|_p = p^{-k},$$

$$\rho_p(x_k, z_k) = \left\| \sum_{i=k+1}^{2k} a_i p^{i-1} \right\|_p = \left\| \left(\sum_{i=t}^{k} a_{i+k} p^{i-1} \right) p^{k+t} \right\|_p = p^{-(k+t)},$$

where a_{k+t} is the first non-zero digit of a_{i+k}. Passing to the limit as $k \to \infty$, prove that the sequences $\{y_k\}$ and $\{z_k\}$ are equivalent to $\{x_k\}$ with respect to the relation φ_p. $\square$

By Definition 6.11, the p-adic numbers are the elements of the factor-set F_p/φ_p that are equivalent fundamental sequences of rational numbers. Particularly, all sequences of Example 6.16 determine the same p-adic number x. Using the denotation of the previous caption, we can use the short form

$$x = [x_k]_p = [y_k]_p = [z_k]_p,$$

where $[x_k]_p$ is the equivalence class of the set F_p with equivalence φ_p and representation $\{x_k\}$. This object is uniquely determined by all p-ary digits a_k; this can be written formally as the infinite in the left set of digits $...a_k a_{k-1}...a_2 a_1, a_0 a_{-1}...a_{-s}$. This is the presentation of p-adic numbers by Definition 6.8 (see Figure 6.4).

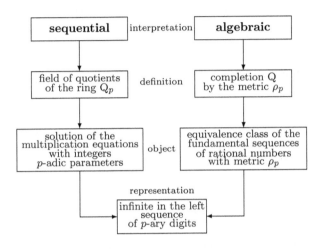

FIGURE 6.4: Algebraic and sequential interpretation of p-adic numbers.

Note that the convergence with respect to the metric ρ_p sequence of rational numbers (for example, the stationary sequence, see Example 6.10) are fundamental and determine the p-adic numbers that are the analogues of the rational real numbers by Cantor interpretation. Thus, the set of rational numbers can be interpreted as a subset of the set of p-adic numbers (see Figure 6.5). However, the divergent fundamental sequences (for example, the sequence $\{x_k\}$ from Example 6.15) can be interpreted as specific analogues of irrational numbers although they naturally have a different nature.

Using the technique of the previous caption, determine the operations on the set of p-adic numbers by Definition 6.11 that is the factor-set F_p/ρ_p. Consider the elements $x = [x_k]_p$ and $y = [y_k]_p$ of this set. Determine its sum and product by the equalities

$$x + y = [x_k]_p + [y_k]_p = [x_k + y_k]_p,$$

$$x \cdot y = [x_k]_p \cdot [y_k]_p = [x_k \cdot y_k]_p.$$

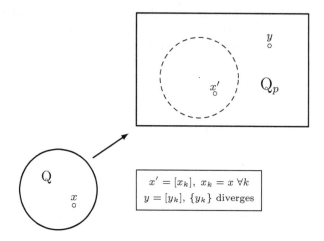

FIGURE 6.5: Sequential extension of the set of rational numbers.

Example 6.17 *Addition p-adic numbers.* Consider the sequence $\{x_k\}$ from Example 6.15, which determine the p-adic number x by Definition 6.11. Determine also the sequence $\{y_k\}$ by the equalities

$$y_k = \sum_{i=t}^{k} b_i p^{i-1}, \; k = 1, 2, \ldots,$$

where t is a natural number, and b_i are p-ary digits. Using the known analysis, determine its fundamentality. Then determine a p-adic number $y = [y_k]_p$ by Definition 5.10 that can be interpreted as the infinite set of p-ary digits $\ldots b_k b_{k-1} \ldots b_2 b_1, b_0 b_{-1} \ldots b_{-t}$. Using the definition of addition, we find the sum

$$x + y = [x_k + y_k]_p.$$

It is based on the sequence of sums $\{x_k + y_k\}$. The element $x_k + y_k$ of this sequence is the result of the bit-by-bit addition of the p-ary number system of the rational numbers x_k and y_k. This is equivalent to the addition in the p-ary number system of the representations $\ldots a_k a_{k-1} \ldots a_2 a_1, a_0 a_{-1} \ldots a_{-s}$ and $\ldots b_k b_{k-1} \ldots b_2 b_1, b_0 b_{-1} \ldots b_{-t}$ of the numbers x and y. However, this is the addition of the p-adic numbers by Definition 6.8 with the same representations. $\square$

Each algebraic operation for the p-adic numbers by Definition 6.8 and Definition 6.11 can be transformed to the analogical algebraic operation for its representations as the infinite in the left of the digits in the p-ary number system (see Figure 6.4). Thus, these definitions give the same algebraic object up to isomorphism. Particularly the set of p-adic numbers by Definition 6.11 with determined operations is the field. We save the same denotation $\mathbb{Q}_p$ for it.

It is possible to determine non-algebraic constructions for the set $\mathbb{Q}_p$ by Definition 6.11. Using the completion method, we determine the metric here as the prolongation of the metric ρ_p. Particularly, for all p-adic numbers $x = [x_k]_p$ and $y = [y_k]_p$ determine the value of the metric (saving its denotation for the rational numbers) by the formula

$$\rho_p(x, y) = \rho_p\big([x_k]_p, [y_k]_p\big) = \lim_{k \to \infty} \rho_p(x_k, y_k).$$

By completion theorem, the set of p-adic numbers is the complete metric space. Besides, the set of rational numbers (more exactly, its isomorphic p-adic analogue) is dense in the set $\mathbb{Q}_p$ with metric ρ_p. Then each p-adic number can be approximated with arbitrary exactness with respect to the metric ρ_p by rational numbers. For example, for all p-adic number $x = [x_k]_p$ can be approximated by the rational number x_k with large enough number k.

Remark 6.17 The set of p-adic numbers does not satisfy the Archimedean axiom.

Thus, there exist different methods of extension of the set of rational numbers to a complete metric space. The result depends upon the choice of the metric. It can be natural metric ρ_∞ or p-adic metric ρ_p (see Figure 6.6).

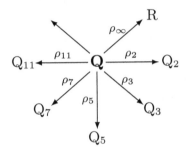

FIGURE 6.6: Sequential extensions of the set of rational numbers.

Remark 6.18 By the Ostrowski theorem, these methods exhaust the possibilities of expanding the set of rational numbers on the basis of the sequential method.

Remark 6.19 There exists the *categories theory*[11] for describing general mathematical constructions. Its influence on mathematics and not only mathematics is steadily increasing. As objects of a category, most often appear as sets, endowed with some specific properties. For example, the groups, the rings, the metric spaces, etc. are the objects of categories. The sets of rational numbers with different metrics that is the pairs $(\mathbb{Q}, \rho_\infty)$, $(\mathbb{Q}, \rho_2)$, $(\mathbb{Q}, \rho_3)$, $(\mathbb{Q}, \rho_5)$,... are the different objects of the category of metric spaces. The transition from one category to another is done with the help of *functors*. The completion theorem actually describes the *completion functor* that maps the category of metric spaces to the category of complete metric spaces. Particularly, this functor transforms the non-complete spaces $(\mathbb{Q}, \rho_\infty)$ and $(\mathbb{Q}, \rho_p)$ to the complete spaces of real and p-adic numbers.

6.6   Conclusions

1. The justification of the determination of the mathematical models is based on the passage to the limit.

2. The effective proof of the convergence is based on the Cauchy criterion that uses the fundamentality of the sequences and does not require a priori knowledge of the limit.

3. The Cauchy criterion is applicable for the complete spaces only.

4. By the completion theorem, each metric space can be extended to a complete metric space such that all elements of this extension can be approximated by elements of the initial space.

5. Each fundamental sequence of the arbitrary space is convergent on the completion of the given space that is the basis of the sequential method.

6. The natural example of using the sequential method is the definition of real numbers as limits of the fundamental sequences of rational numbers.

7. Each real number can be approximated by rational numbers.

8. The set of rational numbers has a different interpretation that is equivalent to the sequential one.

9. Another sequential extension of the set of rational numbers is the class of p-adic numbers.

10. Each p-adic number can be approximated by rational numbers.

11. The set of p-adic numbers has a different interpretation that is equivalent to the sequential one.

We would like to consider also other applications of the sequential method before using it for the justification of mathematical modelling.

Notes

[1]The operations on the residue classes that are the **modular arithmetic** were developed by *Carl Friedrich Gauss* in 1801, see also [10], [117], [120].

[2]The ring is an extension of the field. The theory of rings is described, for example, in [9], [25], [105], [104], [174], [191].

[3]The factor-object in the notion of the theory of categories, see [31], [67], [112], [175].

[4]The general definition of the **isomorphism** is given on the basis of the categories theory, see [31], [67], [112], [175].

[5]p-adic numbers were first described by *Kurt Hensel* in 1897. The definition and the properties of the p-adic numbers are considered in [23], [85], [89], [120], [171].

[6]The **Laurent series** of a complex function is a representation of that function as a power series which includes terms of negative degree. The Laurent series was named after and first published by *Pierre Alphonse Laurent* in 1843 although *Karl Weierstrass* may have discovered it in 1841. The Laurent series is considered in [172], [187].

[7]The ultrametric and the ultrametric space were proposed by *Mark Krasner* in 1944.

[8]The **Ostrowski theorem** was proved by *Alexander Ostrowski* in 1916. This is considered in [89].

[9]The **Diophantine equation** is a polynomial equation, usually in two or more unknowns, such that only the integer solutions are studied, see [120], [122], [129], [155].

[10]The applications of p-adic numbers to theoretical physics are considered in [194], [193], [197].

[11]The categories theory is considered in [31], [67], [112], [175].

Chapter 7

Sequential controls

The determination of mathematical models is connected to the passage to the limit. The Cauchy principle is the practical method of proving of the convergence for the real numbers sequence. However, fundamental sequences may diverge for the incomplete spaces. These spaces can be extended by the completion technique. Then the fundamental sequence becomes convergent on completion. Besides, any element of the extended set can be obtained as a limit of the sequence of elements of the original set. The completion procedure, which is the basis of the sequential method, was tested on an incomplete space of rational numbers. Depending on the choice of the metric on this set, real or p-adic numbers were obtained.

In this chapter we consider sequential objects in an area that does not seem to have any relation to the problems we are discussing. We are talking about the theory of extremum. We consider, in particular, the problem of optimal control, which has no solution in the natural sense. There is no such admissible control on which the least lower bound of the minimized functional is realized here. However, this least lower bound exists, and therefore there is a sequence of controls such that the corresponding values of the functional tends to the least lower bound of this functional. These sequences can be interpreted as in some sense fundamental. However, the absence of optimal control is a sign of its divergence. Therefore, one can try to analyze such problems by expanding the initial set of admissible controls using the completion procedure, i.e. on the basis of the sequential method. In this case, the classes in a certain sense of equivalent sequences of usual controls are interpreted in a natural way as sequential controls.

It is important that each sequential control can be approximated by usual controls. The given functional extends from the set of admissible controls to

155

the set of sequential controls. As a result, we obtain an extended optimal control problem, which is certainly solvable. Moreover, the minimum of the extended functional on the set of sequential controls is equal to the least lower bound of the original functional on a given set of admissible controls. Therefore, we can find an approximate solution of the initial problem under conditions when its exact solution does not exist by the analysis of the extended problem. In addition, we consider the non-uniqueness of the optimal control and the well-posedness of optimization problems in the sense of Tikhonov, which clarifies the structure of sequentially optimal controls.

7.1 Optimal control problems

Optimal control problems often arise in different applications. The object of investigation here is a certain system that can be in different states. Changing a certain factor called the control, we can get this or that state of the system. Note that the possibilities for varying the control are, as a rule, limited. The optimal control problem is to choose a control within the given constraints in order to achieve a goal set.

The problem statement of optimization includes, at first, a mathematical model of the system. Using this model, one can determine a state system for each choice of the control. The mathematical model of the system can be described for the general case by an operator equation

$$A(u, x) = 0, \tag{7.1}$$

which is called the **state equation**. Here u is the **control**, x is the **state function** of the system, and A is an operator defined on the set of control-state pairs. The given constraints can be characterized by the inclusion

$$(u, x) \in U,$$

where U is a given set. One often has constraints for the control only. We have the inclusion $u \in U$ for this case; and U is called the **set of admissible controls**. Finally, a functional I on the set of control-state pairs is given. This is called the **optimality criterion**. Now we give the problem statement of the general optimal control problem[1].

Definition 7.1 *The **optimal control problem** consists of finding a control-state pair that minimizes the functional I on the set U.*

We have the question, why would we consider the problems of finding an extremum? Indeed, this direction is quite far from the main subject of our research. This is the problem of justifying the determination of mathematical

TABLE 7.1: The analogy between extremum theory and equation theory.

minimized object	point	equation	class of equations
function of one variable	number	stationary condition	algebraic equation
function of many variables	vector	vector stationary condition	system of algebraic equations
integral functional (Lagrange problem)	function of one variable	Euler equation	ordinary differential equation
integral functional (Dirichlet integral)	function of many variables	Laplace equation	partial differential equation

physics problems. Is it only the case that here we will encounter sequential control that is another sequential object? However, there exists a deep connection between the theories of the equations and the extremum problems. In particular, the necessary condition of extremum for the function (a minimization object) at some point is the equality to zero of the function derivative at this point. This equality, called the **stationary condition**[2], is an algebraic equation with respect to the suspected number of an extremum. The necessary condition of extremum for the function of many variables is the equality to zero of the gradient of this function. This is the system of algebraic equations. The classic **Lagrange problem** of the **calculus of variations**[3] is related to the minimization of the integral functional, depending on the unknown function and its first derivative, on the set of functions that have fixed values on the boundaries of a given interval. It reduces to a boundary value problem for the **Euler equation**. This is a second order ordinary differential equation. Thus, we have a boundary problem of exactly the same nature as the classical mathematical model of stationary heat transfer considered by us. Finally, the minimization of the integral functional that depends on the function of many variables and its first partial derivatives can be transformed to a second order partial differential equation. For example, the minimization of the **Dirichlet integral**, which in the simplest case is the integral of the sum of the squares of all first derivatives of the function under consideration, reduces to the **Laplace equation**, which consists in zeroing the sum of all the second partial derivatives of the given function. One determines an analogical equation when mathematical modeling of stationary heat transfer in the multidimensional case. The existence of a deep analogy between extremum problems and the theory of equations (see Table 7.1) is a weighty argument in favor of the expediency of considering the sequential method in the theory of optimal control.

Consider a typical example of an optimal control problem.

Example 7.1 Let a system be described by the Cauchy problem

$$\dot{x}(t) = u(t), \quad x(0) = 0 \tag{7.2}$$

that is the partial case of the equation (7.1). The function $u = u(t)$ is the control here. It belongs to the set of square integrable functions on the unit interval such that its values are not greater than 1 for all points. Thus, the set of admissible controls is described by the equality

$$U = \left\{ u \in L_2(0,1) \middle| \; |u(t)| \le 1, \; t \in (0,1) \right\}.$$

The optimal control consists in finding a function u that minimizes the functional

$$I = \int_0^1 (x^2 + u^2) dt$$

on the set U, where $x = x(t)$ is the solution of the system (7.2) for the control u.

This problem is easy enough. It is obvious that the given functional is non-negative. It can be equal to zero whenever the control and the state of the system are equal to zero. The zero control is admissible; and the corresponding solution x of the problem (7.2) is equal to zero too. Thus, this control is the unique solution of the considered optimal control problem, i.e. the **optimal control**. $\square$

Remark 7.1 This optimal control problem has a very good extra property, which is called Tikhonov well-posedness (see Section 7.7).

Remark 7.2 There exist the general solvability theorems for extremum problems[4].

However, there exist extremum problems with qualitatively different properties.

7.2   Insolvable optimal control problems

Consider, at first, the easiest problem of function minimization.

Example 7.2 *Square function on the set of positive numbers.* Consider the problem of minimization for the square function $I(u) = u^2$ on the set U of positive numbers. Its solution does not exist, because of the absence of a minimal positive number. However, the least lower bound of the function I on the given set is equal to zero. This value does not realize on the elements of the set U, but there exists a minimizing sequence $\{u_k\}$ that is a sequence of elements of this set such that the corresponding values of the minimized function tend to its least lower bound on the given set. For example, the numbers $u_k = 1/k$, $k = 1, 2, \dots$ belong to the set U; and the value $I(u_k)$ tends to zero. $\square$

This result seems an analogue of divergent fundamental sequences of the non-complete spaces. The considered sequence is fundamental and divergent on the metric subspace U of the space of real numbers (see Chapter 4). Therefore, we can suppose that these difficulties can be overcome by means of the completion technique.

It is important that there exist elements of the given set such that the corresponding values of the minimized function are close enough to its least lower bound although the minimum of this function does not exist. Therefore, we could try to find the minimizing sequences for determining an approximate solution of the problem. Besides, there is no sense in distinguishing the sequences with the same limit of the values of the minimized function. These sequences are equal in rights by the extremum theory, for example, the, sequence $\{v_k\}$, where $v_k = 1/k^2$, is minimizing and equivalent to the sequence $\{u_k\}$. These sequences can be identified by the factorization. Thus, we consider the equivalence classes of the sequences of the initial set. There are analogues of the elements of the completion of the metric spaces and real and p-adic numbers.

We considered the trivial function on the open set. However, we can obtain analogical results for optimal control problems that are close enough to the problem of Example 7.1.

Example 7.3 *Insolvable optimal control problem*[5]. Consider the following optimal control problem. Suppose a control system is described by the Cauchy problem

$$\dot{x}(t) = u(t), \quad x(0) = 0 \tag{7.3}$$

The control $u = u(t)$ here belongs to the set

$$U = \left\{ u \in L_2(0,1) \middle| \ |u(t)| \leq 1, \ t \in (0,1) \right\}.$$

We would like to find an element u of the set U that minimizes on this set the functional

$$I = \int_0^1 (x^2 - u^2)\,dt,$$

where $x = x(t)$ is the solution of the problem (7.3) for the control u. This problem differs from the problem of Example 7.1 by the sign before the square of the control under the minimized functional.

It is obvious that the value under the given integral is not less than 1. Therefore, we have the inequality $I(u) \geq -1$ for any admissible control u. Consider the sequence of controls determined by the equality (see Figure 7.1):

$$u_k(t) = \begin{cases} 1, & \text{if } \frac{2j}{2k} \leq t < \frac{2j+1}{2k}, \\[2mm] -1, & \text{if } \frac{2j+1}{2k} \leq t < \frac{2j+2}{2k}. \end{cases}$$

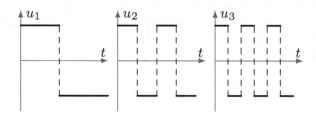

FIGURE 7.1: Sequence of controls.

Consider the corresponding sequence of the states $\{x_k\}$ of the problem (7.3) (see Figure 7.2). For $2j/2k \leq t < (2j+1)/2k$ we have

$$x_k(t) = \int_0^t u_k(\tau)d\tau = \sum_{i=0}^{j-1}\left[\int_{2i/2k}^{(2i+1)/2k} u_k(\tau)d\tau + \int_{(2i+1)/2k}^{(2i+2)/2k} u_k(\tau)d\tau\right] +$$

$$\int_{2j/2k}^t u_k(\tau)d\tau = \sum_{i=0}^{j-1}\left(\frac{1}{2k} - \frac{1}{2k}\right) + \left(t - \frac{2j}{2k}\right) = t - \frac{2j}{2k}.$$

Analogically, for $(2j+1)/2k \leq t < (2j+2)/2k$ we get

$$x_k(t) = \frac{1}{2k} - \left(\frac{2j+1}{2k} - t\right) = \frac{2j+2}{2k} - t.$$

Determine the following inequality (see Figure 7.3)

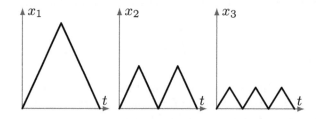

FIGURE 7.2: Sequence of states.

$$0 \leq x_k(t) \leq \frac{1}{2k}, \ t \in (0,1), \ k = 1, 2, \dots .$$

Then we have

$$-1 \le I(u_k) = \int_0^1 (x_k^2 - u_k^2)dt \le \frac{1}{4k^2} - 1, \ k = 1, 2, \dots .$$

Pass to the limit as $k \to \infty$. We have $I(u_k) \to -1$. Thus, the value of

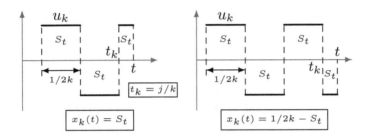

FIGURE 7.3: The value of x_k is not greater than $1/2k$.

the given functional for all admissible controls is not less than -1. However, there exists a sequence of admissible controls such that the corresponding sequence of functionals tends to -1. Therefore, the least lower bound of this functional on the set of admissible control is -1; and the considered sequence is minimizing.

Suppose there exists an admissible control u with value -1 of the functional. It is possible if both following inequalities hold

$$\int_0^1 u^2 dt = 1, \quad \int_0^1 x^2 dt = 0, \tag{7.4}$$

where x is the solution of the problem (7.3) for the control u. From the first equality (7.4) it follows that u is a zero function. Putting it to the equation (7.4), determine that x is zero function too. However, this contradicts the second equality (7.4). Therefore, two equalities (7.4) cannot be true together. Thus, our supposition about the existence of the optimal control for the considered optimization problem is false. $\square$

Thus, this optimal control problem can be insolvable[6].

Remark 7.3 The simplicity and naturalness of this example suggests that the absence of optimal control is quite common.

Remark 7.4 The optimal control problems of Example 7.1 and Example 7.3 have the same state equations and the set of admissible controls. The unique difference here is the sign of one of the terms in the integrand of the minimized functional. This sign affects the extremely important property of a functional, called convexity. Particularly, the functional $I = I(u)$ on a linear space is called **convex**, if for all arguments u and v and any number σ from the unit interval the following inequality holds

$$I[(1 - \sigma)u + \sigma v] \leq (1 - \sigma)I(u) + \sigma I(v).$$

This condition has easy geometric sense. The segment joining any two points on the curve lies no lower than the arc of the curve connecting these points (see Figure 7.4). If it always lies above this arc, then the function is called **strictly convex**. The functional is strictly convex, if the equality for the last relation can be realized for the cases $\sigma = 0$, $\sigma = 1$, and $u = v$ only. One can prove the functional of the Example 7.1 is strictly convex (it has, in reality, a stronger property, see Section 7.7). However, the functional of Example 7.3 is non-convex.

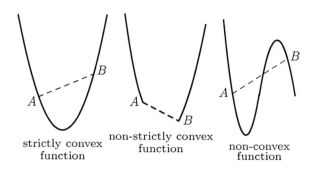

strictly convex   non-strictly convex   non-convex
function   function   function

FIGURE 7.4: Convexity of functions.

Note that the optimal control problem makes sense for the case of its insolvability too, because there exists the least lower bound of the considered functional on the set of admissible controls. Therefore, we can try to find an admissible control such that the corresponding value of the given functional is close enough to this lower bound. However, the classic optimization methods are not applicable here. We try to use here the sequential method.

7.3   Sequential controls

Consider the problem of minimization of the functional I on the control set U. Suppose this functional is lower bounded. Therefore, there exists a lower bound of the given functional on the set U. Then there exists a sequence of elements of this set such that the corresponding sequence of the functional tends to this least lower bound. We can try to find the minimizing sequences whether the task has a solution or not[7].

TABLE 7.2:   Sequential objects

initial object	fundamental sequence $\{u_k\}$	equivalence $\{u_k\}\varphi\{v_k\}$	sequential object	approximation of sequential objects
element of metric space	$\rho(u_{k+r}, u_k) \to 0$	$\rho(u_k, v_k) \to 0$	element of completion	by elements of the initial space
rational number	$\lvert u_{k+r} - u_k \rvert \to 0$	$\lvert u_k - v_k \rvert \to 0$	real number	by rational numbers
rational number	$\lVert u_{k+r} - u_k \rVert_p \to 0$	$\lVert u_k - v_k \rVert_p \to 0$	p-adic number	by rational numbers
admissible control	$\exists \lim I(u_k)$	$[I(u_k) - I(v_k)] \to 0$	sequential control	by usual controls

Consider a set F of the sequences on the set U such that the corresponding sequences of the functional U are convergent, i.e.

$$F = \left\{ \{u_k\} \in U \mid \exists \lim_{k\to\infty} I(u_k) \right\}.$$

It will be the fundamental sequences of the admissible controls. Determine the relation φ such that the condition $\{u_k\}\varphi\{v_k\}$ is true if the limits of these sequences are equal, i.e.

$$\lim_{k\to\infty} I(u_k) = \lim_{k\to\infty} I(v_k).$$

It is obvious, that φ is the equivalence on the set U.

Remark 7.5 The definition of fundamentality does not use the properties of uniform spaces. However, we cannot possibly avoid these properties for the determination of the equivalence of these fundamental sequences. We shall have the analogical situation in connection with definition of sequential distributions, see Chapter 8.

Definition 7.2 *The elements of the factor-set* $V = F/\varphi$ *are called the **sequential controls**[8].*

The sequential controls are the equivalence classes of the fundamental sequences of usual controls. There are the analogues of the elements of the completion of the metric spaces (the equivalence classes of the fundamental sequences of the initial metric space) and real and p-adic numbers (the equivalence classes of the fundamental sequences of rational numbers), see Table 7.2. This predetermines further analysis.

For any control u we can determine the stationary sequence $\{u_k\}$ with element u. The corresponding sequence of functionals $\{I(u_k)\}$ is stationary too, because all its elements are equal to $I(u)$. Then this sequence is convergent, i.e. $\{u_k\} \in F$. Determine the operator $A : U \to V$ that maps a control u to the sequential control Au that is equivalence class $[u_k]$, i.e. the set of all sequences of controls that are equivalent to $\{u_k\}$ (see Figure 7.5). All corresponding sequences of functionals tend to $I(u)$.

Definition 7.3 *The element of the set* (U) *is called the **regular sequential control**, and the element of its complement in V is called the **singular sequential control**.*

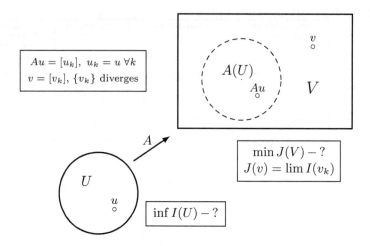

FIGURE 7.5: Sequential extension of extremum problems.

Each regular sequential control contains a stationary sequence of usual controls. Therefore, this is associated with a concrete element of the set of admissible controls.

Remark 7.6 There is no one-to-one correspondence between the sets of usual and regular sequential controls, because the values of the functional at the different controls can be equal (see also Section 7.6 and Section 7.7).

The singular sequential control does not correspond to any elements of the set U. It is obvious that the convergence of the sequence of functionals that is a numerical sequence can be realized for the case of divergent sequence of controls (see, in particular, the minimizing sequence of the considered example). Therefore, the sequential controls are usually singular as the typical real numbers are irrational. Moreover, the set of sequential controls is in some sense the completion of the given set of controls.

Now consider a problem of minimization of a functional on the set V of the sequential controls. Determine the functional

$$J(v) = \lim_{k \to \infty} I(v_k) = \inf I(U).$$

Each sequential control v is the set of all sequences of admissible control $\{v_k\}$ such that the sequences $\{I(v_k)\}$ have the same limit. This is the value of the functional J at the point v of the set V. Besides, the following equality holds

$$I(v) = J(Av) \ \forall v \in U.$$

Remark 7.7 We determine here the prolongation of the given functional from the set of usual controls to the set of sequential controls. These are the analogues of the prolongation of the operations, the metric, and the order (see Chapter 4 and Chapter 6).

Now we have the following solvable extremum problem.

Definition 7.4 *The **extended optimal control problem** is the problem of minimization of the functional J on the set V. Its solution is called the **sequentially optimal control**.*

We have the following relation between usual and sequentially optimal control problems.

Theorem 7.1 *The sequential control v is the sequentially optimal control if and only if each sequence $\{v_k\}$ that determines v is minimizing for the functional I on the set U, besides $\min J(V) = \inf I(U)$.*

Proof. Let $\{v_k\}$ be a sequence on the set U such that $I(v_k) \to \inf I(U)$. Determine the sequential control $v = [v_k]$. We have

$$J(v) = \lim_{k\to\infty} I(v_k) = \inf I(U).$$

If v is not sequentially optimal control, then there exists an element w from the set V such that $J(w) < J(v)$. For any sequence $\{w_k\}$ from w we have

$$\lim_{k\to\infty} I(w_k) = J(w) < J(v) = \inf I(U).$$

Thus, there exists a sequence of the elements of the set U such that the corresponding values of the functional I tend to the least lower bound of this functional on the set of admissible controls. From this contradiction it follows that our supposition about non-optimality of the sequential control v is false.

Now let v be the sequentially optimal control, and $\{v_k\}$ be an arbitrary sequence of the class v. Suppose this is not the minimizing sequence for the functional I on the set U. Then there exists a sequence $\{w_k\}$ of usual controls such that

$$\lim_{k\to\infty} I(w_k) < \lim_{k\to\infty} I(v_k).$$

Determine the sequential control $w = [w_k]$. We obtain the inequality $J(w) < J(v)$. Then v is not sequentially optimal control. Therefore, each sequence of the class v minimizes the functional I on the set U. This completes the proof of the Theorem 7.1. $\square$

Remark 7.8 There exists a different form of extension of the optimal control problem with analogues of Theorem 7.1[9].

By Theorem 7.1, the sequentially optimal control is the equivalence class of all minimizing sequences for the functional I on the set U. The extended extremum problem is solvable if the initial functional is lower bounded on the set of admissible control.

Theorem 7.2 *The optimal control problem is solvable if and only if the corresponding sequentially optimal control is regular.*

Proof. If there exists an optimal control u for the problem of minimization of the functional I on the set U, then we have the equality $J(u) = I(u)$. Using Theorem 7.1, determine that the regular sequential control Au is sequentially optimal. If the sequentially optimal control is regular, then it is equal to Au for a usual control u, besides $I(u) = J(u)$. By Theorem 7.1, the value of the functional I at the point u is equal to its least lower bound on the set U. Therefore, u is the solution of the initial optimal control problem. $\square$

Remark 7.9 We can interpret the insolvability of the optimization problem as the special form of non-completeness of the space.

Remark 7.10 The properties of the sequential optimal control will be exactly specified in Section 7.6 and Section 7.7.

Remark 7.11 Solution of the extended control problem, i.e. the sequentially optimal control can be interpreted as a generalized solution of the initial optimal control problem. Then its classic solution is the usual optimal control. If the given problem is solvable, then the sequentially optimal control is regular. However, the sequentially optimal control, i.e. the generalized solution of the initial optimal control problem, can exist in the case of the absence of its classic solution. Therefore, we have the analogical relation between classic and generalized solutions of the extremum problem and the boundary problem for differential equations (see Chapter 2).

Remark 7.12 It seems more exact to interpret the sequentially optimal control as the sequential solution of the given optimal control problem. Its analogue for the mathematical physics problem will be determined in Chapter 8.

The optimal control problem is insolvable if the sequentially optimal control is singular. Typical of the singularity of the sequential controls, the extremum problems of general form, as a rule, have no solution. In itself, this fact is well known. However, using sequential controls has a natural explanation: regular sequential controls are too small to cover sequentially optimal controls for general problems.

It is very important that the insolvable problem has a sense (see Figure 7.6). Particularly, we can try to find an approximate solution of the problem. This is an admissible control such that the corresponding value of the minimized functional is close enough to its least lower bound on the set of admissible control. The transition from finding the nonexistent optimal control to finding the existing sequentially optimal control can give a basis for determining the minimizing sequences that generate it. Then the element of a minimizing sequence with large enough number can be chosen as the approximate solution of the given optimal control problem. The possibility of the approximation of the sequential object by usual objects (see Table 7.2) is the basis of the sequential method. Therefore, we determine the usual control (the approximate solution of the optimal control problem) as the result, just as practical work with an irrational numbers reduces to working with their rational approximation.

Remark 7.13 We can have a different form of approximate solution of optimal control problems. Let us have the problem of the minimization of a functional I on a subset U of a metric space V. The most natural is the strong form of the approximate solution of the problem. The control u from the set U is called the ***strongly approximate*** (more exact, ω-strong) ***solution*** of the considered problem, if the following inequality holds

$$\rho\big(u, u_{opt}\big) \leq \omega$$

for a small enough positive number ω, where u_{opt} is the exact solution of the problem, and ρ is the metric of the space V. Of course, for the insolvable problems this form of approximate solution does not make any sense. However, we can determine its other form. The control u from the set U is called the ***weakly approximate*** (more exact, ε-weak) ***solution*** of the considered problem, if the following inequality holds

$$I(u) \leq \inf I(U) + \varepsilon$$

for a small enough positive number ε. If the functional I is continuous, then each strongly approximate solution of the problem is its weakly approximate solution. The weakly approximate solutions make sense for insolvable optimal control problems too. Then we can try to find them.

Remark 7.14 There exists a possibility to weaken the notion of the approximate solution of optimal control problems. Consider the general problem of Remark 7.13. The control u from the set U is called the ***weakened approximate*** (more exact, (ε, δ)-weakened) ***solution*** of the considered problem, if the following inequalities hold

$$I(u) \leq \inf I(U) + \varepsilon \quad \rho(u, U) \leq \delta,$$

for small enough positive numbers ε, δ, where

$$\rho(u, U) = \inf_{v \in U} \rho(u, v).$$

One can ask whether it is permissible to use weakened approximate, which can be non-admissible control. However, we use the non-optimal strong approximate solution, because this is close enough to optimal control. We use also the weak approximate solution, which can be far from optimal control (see, particularly, Example 7.7), because the value of the minimized functional at this point is close enough to its least lower bound on the set of admissible control. Then we can use the weakened approximate solution too, because it is close enough to an admissible control, and the value of the minimized functional at this point is close enough to its least lower bound on the set of admissible control. Obviously, each weakly approximate solution of the problem is its weakened approximate solution. This is the weaker notion. However, the class of applicability of the weakened approximate solution is larger[10]. The relation between different forms of approximate solutions of optimal control problems is presented in Table 7.3.

Remark 7.15 One can determine the sequential extension of the optimal control problem, which corresponds to the weakened approximate solution of optimal control problems. Particularly, the set F (see definition of sequential controls) can by determined by the formula

$$F = \Big\{ \{u_k\} \in V \Big| \; \exists \lim_{k \to \infty} I(u_k); \;\; \rho(u_k, U) \to 0 \Big\}.$$

Then we could determine sequential controls and proof the analogue of Theorem 7.1.

The question arises, how can we in practice characterize sequentially optimal control and determine the minimizing sequence that generates it? Consider, at first, the easiest insolvable problem of function minimization (see Example 7.2).

TABLE 7.3: Different forms of approximate solutions of optimal control problems

approximate solutions	control admissibility	functional closeness	control closeness
ω-strong	yes	yes	yes
ε-weak	yes	yes	no
(ε, δ)-weakened	no	yes	no

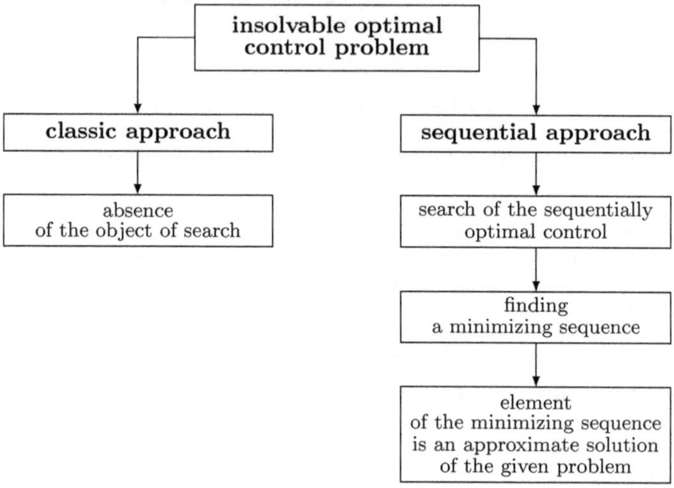

FIGURE 7.6: Analysis of the insolvable optimal control problem.

7.4 Extension of the easiest extremum problem

The existence of the solution of the extremum problem is necessary for using the classic optimization methods. Therefore, the analysis of the insolvable optimal control problem is difficult enough. However, if we find an existing solution of the extended problem, then we can try to determine a minimizing sequence for the initial problem for the case of its insolvability too. This method seems very difficult, because it is necessary to find a sequential control that is the set of the sequences of the usual controls. However, we are working with real numbers that are the classes of sequences of rational numbers by Cantor's interpretation.

Determine, at first, the sequential extension of the minimization problem for the square function on the set of positive numbers (see Example 7.2).

Example 7.4 Consider the minimization problem of the square function $I(u) = u^2$ on the set of positive numbers U. For any sequence $\{u_k\}$ of the set

U the convergence of the sequence $\{I(u_k)\}$ is realized whenever the sequence $\{u_k\}$ is convergent. Then the set F consists of all sequences of positive numbers that are convergent on the set of real numbers. The sequences $\{u_k\}$ and $\{v_k\}$ of the set F are equivalent, if

$$\lim_{k\to\infty} I(u_k) = \lim_{k\to\infty} I(v_k).$$

This is true for the case of the equality of the limits for the considered sequences. Therefore, each sequence $\{u_k\}$ determines the sequential control $[u_k]$ that is the set of all sequences of the set U with same limit as $\{u_k\}$. Thus, the sequential controls are the set of equivalence classes of the sequences of positive numbers that are convergent on the metric set of real numbers. This is the completion of the space of positive numbers.

If the sequence $\{u_k\}$ has a limit u on the set of positive numbers, then it is equivalent to the stationary sequence with element u. This sequential control is regular and can be identified with the usual control u that is a number. If a fundamental sequence of the set U is divergent that can be possible if it tends to zero, then the corresponding sequential control is singular. It can be identified with the number 0 that is not the element of the set U. Thus, the set of sequential controls V coincides up to isomorphism with the set of non-negative numbers; besides any positive number is regular, and the number 0 is singular.

Determine the functional

$$J(u) = \lim_{k\to\infty} I(u_k) = \lim_{k\to\infty} (u_k)^2 \ \forall\{u_k\} \in u$$

on the set V. By Theorem 7.1, each minimizing sequence determines a sequential control that minimizes the functional J on the set V. The sequentially optimal control exists here because of the boundedness of the function I on the set U.

Determine **necessary conditions of extremum** for the extended problem. There are the relations that the solution of the extremum problem must satisfy. We use an algebraic property of the set U. The subset of a linear space is called **convex**, if for all its elements u and v the object $\sigma u + (1\sigma)v$ belongs to this set for any number σ from the interval $[0,1]$ (see Figure 7.7).

Suppose one has sequential controls v_1 and v_2 that are determined by the sequences $\{v_{1k}\}$ and $\{v_{2k}\}$ from U. By the convexity of the set U, the point

$$w_k = \sigma v_{1k} + (1-\sigma)v_{2k}$$

belongs to this set for all $\sigma \in (0,1)$. Then the object $\sigma v_1 + (1-\sigma)v_2$ is the element of the set V that is determined by the sequence $\{w_k\}$. The following inequality holds

$$J[u + \sigma(v-u)] - J(u) \geq 0 \ \forall v \in V, \sigma \in (0,1).$$

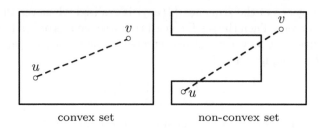

convex set   non-convex set

FIGURE 7.7: Convex set includes the segment connecting any of its points.

It can be transformed to

$$\lim_{k \to \infty} \left[u_k + \sigma(v_k - u_k) \right]^2 - \lim_{k \to \infty} (u_k)^2 \geq 0 \ \forall v \in V, \ \sigma \in (0,1).$$

The numerical sequences $\{u_k\}$ and $\{v_k\}$ belong to the set F. Therefore, it has limits u' and v'. From the last inequality, it follows that

$$2\sigma u'(v' - u') + \sigma^2 (v' - u')^2 \geq 0.$$

Dividing by σ and passing to the limit as $\sigma \to 0$, we obtain the inequality

$$\sigma u'(v' - u') \geq 0.$$

This is true for any element v' that is the limit of the arbitrary convergent sequence of positive numbers. The set of these limits is the set of non-negative numbers. Then we get

$$u'(v' - u') \geq 0 \ \forall v' \geq 0.$$

This relation is called the ***variational inequality***[11]. Its unique solution is $u' = 0$. Thus, the sequence $\{u_k\}$ that determines the sequentially optimal control tends to zero. By Theorem 7.1, this sequence is minimizing for the given extremum problem. □

This example illustrates the method of analysis for insolvable problems by the extension idea. The minimization problem of the square function on the set of positive numbers does not have any solutions. However, there exists the least lower bound of this function on the given set. Then there exists an element of the set U such that the value of the considered function there is close enough to this least lower bound. We can find it, if we determine a minimizing sequence. The direct use of the necessary condition of extremum here is not applicable because of the absence of the solution for the given problem. However, we can find a minimizing sequence by the analysis of the extended extremum problem that is solvable.

7.5   Extension of the optimal control problem

Apply the previous method for the analysis of the insolvable optimal control problem (see Example 7.3). By the Bolzano–Weierstrass theorem, for all bounded sequences of real numbers there exists a convergent subsequence (see Chapter 5). Its generalization follows the **Banach–Alaoglu theorem**[12].

Theorem 7.3 *For all bounded with respect to the norm sequence of the Hilbert space there exists a weakly convergent subsequence.*

Remark 7.16 In reality, the Banach–Alaoglu theorem is a stronger assertion. However, Theorem 7.3, which is its weaker form, will be sufficient for us[13].

We consider the **Sobolev space** $H^1(0,1)$ of the square integrable with its first derivatives functions on the unit interval. This is the Hilbert space with scalar product

$$(x,y) = \int\limits_0^1 \left[x(t)y(t) + \frac{dx(t)}{dt}\frac{dy(t)}{dt} \right] dt.$$

Consider the **Rellich–Kondrashov theorem**[14].

Theorem 7.4 *From the convergence $x_k \to x$ weakly in $H^1(0,1)$, it follows that $x_k \to x$ strongly in $L_2(0,1)$.*

Remark 7.17 In reality, the Rellich–Kondrashov theorem is a stronger assertion. However, Theorem 7.4 will be sufficient for us. The sense of this theorem obtains the strong convergence in the "weaker" space from the weak convergence in the "stronger" space.

Note also the **Schwartz inequality**

$$|(x,y)| \leq \|x\|\|y\|;$$

i.e. the absolute value of the scalar product is not greater than the product of norms.

Remark 7.18 These assertions will be used also in Chapter 9 for the analysis of the models of the stationary heat transfer phenomenon.

Now we return to the consideration of Example 7.3.

Example 7.5 We consider of the problem of finding a function u that minimizes the functional

$$I = \int\limits_0^1 (x^2 - u^2) dt$$

on the set

$$U = \left\{ u \in L_2(0,1) \mid |u(t)| \leq 1, \ t \in (0,1) \right\},$$

where x is the solution of the problem

$$\dot{x}(t) = u(t), \quad x(0) = 0.$$

Suppose the sequences $\{u_k\}$ and $\{v_k\}$ of the set U are equivalent with respect to the relation φ, if the following conditions hold

$$(u_k - v_k) \to 0 \text{ in } L_2(0,1), \quad \lim_{k \to \infty} I(u_k) = \lim_{k \to \infty} I(v_k). \tag{7.5}$$

Consider the problem

$$\dot{z}_k = (u_k - v_k), \ t \in (0,1); \ \ z_k(0) = 0.$$

Integrating, we get

$$z_k(t) = \int_0^t \big[u_k(\tau) - v_k(\tau) \big] d\tau.$$

The value at the right-hand side of this equality can be interpreted as the scalar product of the function that equals to 1 everywhere and the function $u_k - v_k$ of the space $L_2(0,t)$. Using the Schwartz inequality, we have

$$\big| z_k(t) \big| \le \left(\int_0^t 1 d\tau \right)^{1/2} \left[\int_0^t \big| u_k(\tau) - v_k(\tau) \big|^2 d\tau \right]^{1/2} \le \sqrt{t} \| u_k - v_k \|$$

with norm of the space $L_2(0,t)$. Using the first condition (7.5), we obtain the convergence $z_k \to 0$ in $L_2(0,t)$. Let x_k and y_k be the state functions for the controls u_k and v_k. By the equality $z_k = x_k - y_k$, we get

$$\| x_k \| = \| y_k + z_k \| \le \| y_k \| + \| z_k \|, \quad \| y_k \| = \| x_k - z_k \| \le \| x_k \| + \| z_k \|.$$

Then we obtain the inequality

$$\big| \| x_k \| - \| y_k \| \big| \le \| z_k \|.$$

From the convergence $\| z_k \| \to 0$, it follows that

$$\lim_{k \to \infty} \| x_k \|^2 = \lim_{k \to \infty} \| y_k \|^2.$$

Using the analogical transformations, we have the equality

$$\lim_{k \to \infty} \| u_k \|^2 = \lim_{k \to \infty} \| v_k \|^2.$$

Our optimality criterion is the difference between the squares of norms of the state function and the control. Therefore, the second condition (7.5) is the corollary of its first condition. Thus, the relation $\{u_k\} \varphi \{v_k\}$ is true whenever

$(u_k - v_k) \to 0$ in $L_2(0,1)$. Then we determine the sequential control as the equivalence class of the sequences

$$[u_k] = \left\{ \{v_k\} \subset U \middle| v_k = u_k + w_k, \ w_k \to 0 \text{ in } L_2(0,1) \right\}.$$

This object is regular if there exists a generating sequence $\{u_k\}$ that is convergent in the space $L_2(0,1)$ and singular if this sequence does not exist. The regular sequential control can be identified with the limit of the convergent generating sequence.

Determine the functional

$$J([u_k]) = \lim_{k \to \infty} I(u_k)$$

on the set V of sequential controls. The extended optimal control problem is the problem of minimization for the functional J on the set V. By the boundedness of the functional I on the set U, the extended problem is solvable.

Consider sequential controls u_1 and u_2. We have the equalities

$$u_i = [u_{ik}] = \left\{ \{v_{ik}\} \subset U \middle| v_{ik} = u_{ik} + w_{ik}, \ w_{ik} \to 0 \text{ in } L_2(0,1) \right\}, \ i = 1,2.$$

Then we get

$$\sigma(u_{1k} + w_{1k}) + (1 - \sigma)(u_{2k} + w_{2k}) \in U, \ k = 1,2,\dots.$$

The sequence $\{w_k\}$, determined by the equalities

$$w_k = \sigma w_{1k} + (1 - \sigma)w_{2k}, \ k = 1,2,\dots,$$

is strongly convergent to zero in $L_2(0,1)$. Thus, the object

$$u = \left[\sigma u_{1k} + (1 - \sigma)u_{2k} \right]$$

is the sequential control. Determine the linear space on the factor-set of sequences of $L_2(0,1)$ with equivalence φ by the standard method (see Chapter 4). We get

$$\sigma u_1 + (1 - \sigma)u_2 = \left[\sigma u_{1k} + (1 - \sigma)u_{2k} \right] = u.$$

Thus, the value at the right-hand side of this equality is the sequential control. Therefore, the set V is convex.

Suppose u is the sequentially optimal control; and v is a sequential control that is determined by the equality

$$v = [v_k] = \left\{ \{v_k + w_k\} \in U \middle| w_k \to 0 \text{ in } L_2(0,1) \right\}.$$

Using the convexity of the set V, determine the inclusion of the sequential control $\sigma v + (1 - \sigma)u$ to the set V. Then we have the inequality

$$J\left[\sigma v + (1 - \sigma)u \right] - J(u) \geq 0, \ \sigma \in (0,1).$$

Now we get

$$\lim_{k\to\infty} I\big[u_k + \sigma(v_k - u_k)\big] - \lim_{k\to\infty} I(u_k) \geq 0. \tag{7.6}$$

Determine the value

$$I\big[u_k + \sigma(v_k - u_k)\big] = \|z_k\|^2 - \big\|u_k + \sigma(v_k - u_k)\big\|^2,$$

where z_k is the solution of the problem

$$\dot{z}_k = u_k + \sigma(v_k - u_k), \ t \in (0,1); \ \ z_k(0) = 0.$$

Denote again by x_k and y_k the state functions for the controls u_k and v_k. Determine the function

$$z_k = x_k + \sigma(y_k - x_k).$$

We have the equality

$$I\big[u_k + \sigma(v_k - u_k)\big] = \|x_k\|^2 - \|u_k\|^2 + 2\sigma\big(x_k, y_k - z_k\big) - 2\sigma\big(u_k, v_k - u_k\big) + \sigma^2 \eta_k$$

with the scalar product of the space $L_2(0,1)$, where

$$\eta_k = \|y_k - x_k\|^2 - \|v_k - u_k\|^2.$$

Consider the problem

$$\dot{p}_k = -x_k, \ t \in (0,1); \ \ p_k(1) = 0 \tag{7.7}$$

that is called the **adjoint system**. Multiply the equality (7.7) by the difference $y_k - x_k$ and integrate the result. Using the formula of integration by part, we get

$$\big(x_k, y_k - x_k\big) = \big(p_k, v_k - u_k\big).$$

Thus, we obtain

$$I\big[u_k + \sigma(v_k - u_k)\big] = \|x_k\|^2 - \|u_k\|^2 + 2\sigma\big(p_k - u_k, v_k - u_k\big) + \sigma^2 \eta_k. \tag{7.8}$$

The sequences $\{u_k\}$ and $\{v_k\}$ are bounded in the space $L_2(0,1)$ because of the boundedness of the set of admissible controls. Using the Banach–Alaoglu theorem, after extracting subsequences we have the convergence $u_k \to u'$ weakly in $L_2(0,1)$, and $v_k \to v'$ weakly in $L_2(0,1)$. Find the solution of the state equation

$$x_k(t) = \int_0^t u_k(\tau)d\tau.$$

Using the boundedness of the set of admissible controls, we have the boundedness of the sequence $\{x_k\}$ in $L_2(0,1)$. The sequence of derivatives $\{\dot{x}_k\}$ is bounded in $L_2(0,1)$ by the state equation. Then the sequence $\{x_k\}$ is bounded in $H^1(0,1)$. Therefore, we obtain the convergence $x_k \to x'$ weakly in $H^1(0,1)$

because of the Banach–Alaoglu theorem. By the equality $\dot{x}_k = u_k$, x' is the state function for the control u'. We prove also the convergence $y_k \to y'$ weakly in $H^1(0,1)$, where y' is the state function for the control v'. By the Rellich–Kondrashov theorem, we have the convergence $x_k \to x'$ and $y_k \to y'$ strongly in $L_2(0,1)$. Using the analogical transformations, we obtain $p_k \to p'$ in $L_2(0,1)$.

Determine the inequality

$$\left| (p_k, v_k - u_k) - (p', v' - u') \right| \le \left| (p_k - p', v_k - u_k) \right| + \left| (p', (v_k - u_k) - (v' - u')) \right| \le$$

$$\|p_k - p'\|\|v_k - u_k\| + \left| (p', v_k - v') \right| + \left| (p', u_k - u') \right|.$$

Then we have the convergence

$$(p_k, v_k - u_k) \to (p', v' - u').$$

From the equality (7.8), it follows that

$$\lim_{k \to \infty} I[u_k + \sigma(v_k - u_k)] = \|x\|^2 - \lim_{k \to \infty} \|u_k\|^2 + 2\sigma(p', v' - u') -$$

$$2\sigma \lim_{k \to \infty} (u_k, v_k) + 2\sigma \lim_{k \to \infty} \|u_k\|^2 + \sigma^2 \lim_{k \to \infty} \eta_k.$$

Using the equality

$$\lim_{k \to \infty} I(u_k) = \|x'\|^2 - \lim_{k \to \infty} \|u_k\|^2,$$

transform the inequality (7.6)

$$(p', v' - u') - \lim_{k \to \infty} (u_k, v_k) + \lim_{k \to \infty} \|u_k\|^2 + \sigma/2 \lim_{k \to \infty} \eta_k \ge 0.$$

Passing to the limit as $\sigma \to 0$ using the boundedness of the sequence $\{\eta_k\}$, we have the inequality

$$(p', v' - u') + \lim_{k \to \infty} \|u_k\|^2 \ge \lim_{k \to \infty} (u_k, v_k). \tag{7.9}$$

This is true for any sequence $\{v_k\}$ of admissible controls. This condition is necessary for the sequential optimality of the sequential control u that is generated by the sequence $\{u_k\}$. By the Schwartz inequality and the definition of the set U, we get

$$\lim_{k \to \infty} (u_k, v_k) \le \lim_{k \to \infty} \|u_k\|\|v_k\| \le 1.$$

Then the condition (7.9) is true if the following inequality holds

$$(p', v' - u') + \lim_{k \to \infty} \|u_k\|^2 \ge 1.$$

The function v' is here the weak limit of the arbitrary sequence $\{v_k\}$ of admissible controls. The second term in the last inequality is not greater than 1. Then the last relation is true, at least, if the following equalities hold

$$p' = 0, \ \lim_{k \to \infty} \|u_k\| = 1.$$

Multiply the adjoint equation (7.7) by a smooth enough function $\lambda = \lambda(t)$ that is equal to zero for $t = 0$. After integration using the formula of integration by parts we get

$$(p_k, \dot{\lambda}) = -(x_k, \lambda).$$

Passing to the limit, we have

$$-(x', \lambda) = (p', \dot{\lambda}) = 0.$$

Then $x' = 0$ because the function λ is arbitrary. After analogical transformations for the state equation we find $u' = 0$. Thus, we obtain the following conditions for the sequence $\{u_k\}$ that determines the sequentially optimal control

$$u_k \to 0 \text{ weakly in } L_2(0,1), \ \|u_k\| \to 1. \tag{7.10}$$

Find the value of the functional J at the sequential control generated by the sequence $\{u_k\}$ that satisfies the condition (7.10). We have the convergence $x_k \to 0$ strongly in $L_2(0,1)$. Hence, $\|x_k\| \to 0$. Then we get

$$J\big([u_k]\big) = \lim_{k \to \infty} \big(\|x_k\|^2 - \|u_k\|^2\big).$$

Passing to the limit using the conditions (7.10), determine the convergence $J\big([u_k]\big) \to -1$. However, the minimum of the functional J on the set V is equal to the least lower bound of the functional I on the set U that is equal to -1. Thus, we describe, in reality, minimizing sequences of the initial optimal control problem. $\square$

Remark 7.19 Each sequence $\{v_k\}$ that is equivalent to the sequence $\{u_k\}$ satisfying the conditions (7.10) is minimizing too. Indeed, each control v_k here is determined by the equality $v_k = u_k + w_k$, where $w_k \to 0$ strongly in $L_2(0,1)$. Using (7.10), we have $v_k \to 0$ weakly in $L_2(0,1)$. From the inequality

$$\big|\|v_k\| - \|u_k\|\big| \le \|w_k\|,$$

it follows that the sequence $\{v_k\}$ satisfies the conditions (7.10) too. Therefore, this is the minimizing sequence. Note that the sequence of Example 7.3 satisfies the conditions (7.10).

Thus, we can determine the class of minimizing sequences. Therefore, we have many admissible controls such that the corresponding values of the minimized functional are close enough to its least lower bound.

Remark 7.20 We cannot guarantee that the necessary conditions of the sequential optimality do have another solutions. Besides, the solution of necessary conditions of optimality can be non-optimal.

Remark 7.21 We could determine the variational inequality

$$(p - u, v - u) \geq 0 \ \forall v \in U$$

as the necessary condition of optimality for the considered problem, where p is the solution of the adjoint system

$$\dot{p} = -x, \ t \in (0, 1); \ p(1) = 0.$$

However, this optimality condition has no solution. Therefore, we cannot use it for finding an approximate solution of the given optimal control problem.

Now we consider additional examples for analysis of the properties of the sequentially optimal control.

7.6 Non-uniqueness of the optimal control

Consider the following optimal control problem[15].

Example 7.6 We have again the state system described by the Cauchy problem

$$\dot{x}(t) = u(t), \ x(0) = 0.$$

The control $u = u(t)$ belongs again to the set

$$U = \left\{ u \in L_2(0, 1) \middle| \ |u(t)| \leq 1, \ t \in (0, 1) \right\}.$$

We have the minimization problem for the functional

$$I = - \int_0^1 (x^2 + u^2) dt$$

on the set U, where is the state function for the control u. This problem has a unique difference from the optimization problems of Example 7.1 and Example 7.3. This is the sign of terms of integrands. However, this optimization problem has qualitatively different properties.

Using the definition of the set of admissible control, we have the inequality

$$-1 \leq u(t) \leq 1, \ t \in (0, 1).$$

Determine the state function

$$x(t) = \int_0^t u(t) dt.$$

Integrating the previous inequality, we have

$$-t \leq x(t) \leq t, \ t \in (0, 1).$$

Then we get

$$0 \leq [u(t)]^2 \leq 1, \ \ 0 \leq [x(t)]^2 \leq t^2.$$

Therefore, the following inequality holds

$$-(1 + t^2) \leq -\{[x(t)]^2 + [u(t)]^2\} \leq 0, \ t \in (0,1).$$

After integration we prove that the value of the given functional at the arbitrary admissible control satisfies the inequality

$$-4/3 \leq I \leq 0.$$

The value $-4/3$ can be realized only if the squares of the state and of the control have their maximum possible values. Therefore, the considered optimal control problem has two solutions

$$u_1(t) = 1, \ \ u_2(t) = -1, \ \ t \in (0,1).$$

Consider the operator A that maps the control u to the corresponding sequential control Au. This is the equivalence class of the sequences that is equivalent to the stationary sequence with element u. We have the equality

$$A(u_1) = A(u_2),$$

because the values of the functional at both controls are the same. Therefore, we have the unique sequential control for both usual optimal controls.

Determine now the sequence of controls $\{u_k\}$ with elements $u_1, u_2, u_1, u_2, u_1, u_2$, etc. The corresponding sequence of the functional values is stationary with element $-4/3$ that is the minimum of the given functional. Note that the divergent control sequence $\{u_k\}$ generates the regular sequential control. This sequential control is generated also by the stationary sequences with element 1 or -1. $\square$

Remark 7.22 One can prove that the optimal control problem with strictly convex functional cannot have more than one solution. The strict convexity of the minimized functional is realized for Example 7.1, not for Example 7.6.

The analogical results are true for the general case too. Particularly, we have the following assertion.

Theorem 7.5 *The sequentially optimal control is unique. If the initial optimal control problem has a non-unique solution, then the corresponding sequentially optimal control can be generated by convergent and divergent sequences.*

We will know that non-trivial properties of the sequentially optimal control can even be realized for the uniqueness of the optimal control.

7.7 Tihonov well-posedness of the optimal control problems

Consider another optimal control problem[16].

Example 7.7 Let U be the set of square integrable functions $u = u(t)$ on the unit interval such that $|u(t)| \leq 1$ on $[0,1]$. We have the problem of minimization of the functional

$$I = \int_0^1 x^2 dt,$$

where $x = x(t)$ is the solution of the Cauchy problem

$$\dot{x}(t) = u(t), \quad x(0) = 0.$$

It is obvious that the given functional is non-negative. It can be equal to zero if the function x is equal to zero everywhere. The state function here is equal to zero for zero control only. Therefore, the given optimal control problem has the unique solution $u_0 = 0$. Besides, the minimum of the functional on the set of admissible controls is equal to zero.

Consider the following sequence (see Figure 7.8)

$$u_k(t) = \sin \pi k t, \quad k = 1, 2, \dots .$$

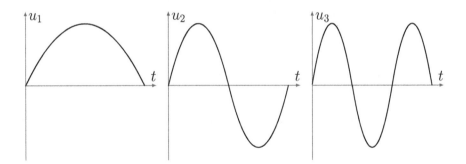

FIGURE 7.8: Minimizing sequence for Example 7.7.

These functions are infinitely differentiable; and the values $|u_k(t)|$ are not greater than 1. Thus, we have the sequence of admissible controls. The corresponding state functions are determined by the formula (see Figure 7.9):

$$x_k(t) = \int_0^t \sin \pi k \tau d\tau = \frac{1 - \cos \pi k t}{\pi k}.$$

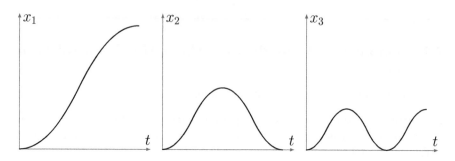

FIGURE 7.9: The sequence of the states for the minimizing sequence.

We have the inequality

$$0 \le x_k(t) \le 2/k\pi, \ t \in (0,1), \ k = 1, 2, \dots .$$

After integration we get

$$0 \le \int_0^1 x_k^2 dt \le \frac{4}{(k\pi)^2}, \ k = 1, 2, \dots .$$

Then the sequence $\{I(u_k)\}$ tends to zero that is the minimum of the given functional on the set of admissible control. Therefore, the considered control sequence is minimizing. Check the convergence of this sequence to the unique optimal control u_0. We find the value

$$\|u_k - u_0\|^2 = \int_0^1 |u_k(t) - u_0(t)|^2 dt = \int_0^1 \sin^2 k\pi t dt = \frac{1}{2}.$$

Thus, the optimal control u_0 is not the limit of the minimizing sequence $\{u_k\}$. However, this sequence determines the sequentially optimal control for the considered example. □

By Example 7.7, the absence of the convergence of the minimizing sequence to the optimal control for the case of the uniqueness of the optimal control is even possible.

Remark 7.23 One can prove that the considered minimizing sequence is divergent in the space $L_2(0,1)$.

Definition 7.5 *The optimal control problem is called **well-posed by Tikhonov**[17], if it is solvable, and each minimizing sequence tends to the optimal control.*

Remark 7.24 If the optimal control problem is well-posed by Tikhonov, then the weak-approximate solution is strong-approximate. Therefore, these notions are equivalent. Of course, this is not true for ill-posed problems.

Remark 7.25 One considers also the optimal control problems that are *well-posed by Hadamard*[18]. However, this property is far from the considered problems.

By Example 7.6, the optimal control problem with non-unique solutions is ill-posed by Tikhonov. The question arises, is there a problem that is well-posed by Tikhonov?

Example 7.8 Return to Example 7.1. It is necessary to choose the control u from the set

$$U = \left\{ u \in L_2(0,1) \mid |u(t)| \leq 1, \ t \in (0,1) \right\}$$

that minimizes there the functional

$$I = \int_0^1 (x^2 + u^2) dt,$$

where $x = x(t)$ is the solution of the system

$$\dot{x}(t) = u(t), \quad x(0) = 0.$$

We know that this problem has the unique $u_0 = 0$; and the corresponding optimal state is $x_0 = 0$.

Denote by $\{u_k\}$ an arbitrary minimizing sequence for the considered optimal control problem. Let $\{x_k\}$ be the corresponding sequence of state functions. Consider the control

$$v_k = \sigma u_k + (1\sigma) u_0$$

that belongs to the set U for any number σ from the interval $[0,1]$. Denote by y_k the solution of the state system for the control v_k. From the equalities

$$\dot{x}_0(t) = u_0(t), \ x_0(0) = 0; \ \dot{x}_k(t) = u_k(t), \ x_k(0) = 0; \ \dot{y}_k(t) = v_k(t), \ y_k(0) = 0$$

it follows that

$$y_k = \sigma x_k + (1 - \sigma) x_0. \tag{7.11}$$

Find the value of the functional

$$I(v_k) = \int_0^1 (y_k^2 + u_k^2) dt = \int_0^1 \left\{ [\sigma x_k + (1 - \sigma) x_0]^2 + [\sigma u_k + (1 - \sigma) u_0]^2 \right\} dt.$$

We have the inequality

$$[\sigma x_k + (1 - \sigma) x_0]^2 \leq \sigma x_k^2 + (1 - \sigma) x_0^2, \tag{7.12}$$

because of the convexity of the square function. Besides, we have the equality

$$\left[\sigma u_k + (1-\sigma)u_0\right]^2 = \sigma^2 u_k^2 + 2\sigma(1-\sigma)u_k u_0 + (1-\sigma)^2 u_0^2.$$

Then we get

$$\left[\sigma u_k + (1-\sigma)u_0\right]^2 = \sigma u_k^2 + (1-\sigma)u_0^2 - 2\sigma(1-\sigma)(u_k - u_0)^2. \qquad (7.13)$$

Adding the relations (7.12) and (7.13), after integration we have the inequality

$$I\left(\sigma u_k + (1-\sigma)u_0\right) \leq \sigma I(u_k) + (1-\sigma)I(u_0) - 2\sigma(1-\sigma)\|u_k - u_0\|^2 \quad (7.14)$$

with norm of the space $L_2(0,1)$. From (7.14), it follows that

$$I\left(\sigma u_k + (1-\sigma)u_0\right) - I(u_0) \leq \sigma\left[I(u_k) - I(u_0) - 2\sigma(1-\sigma)\|u_k - u_0\|^2\right].$$

The value of the left-hand of this inequality is non-negative, because the control u_0 is optimal. Dividing by σ, we get

$$2\sigma(1-\sigma)\|u_k - u_0\|^2 \leq I(u_k) - I(u_0).$$

Pass to the limit here as $k \to \infty$. Then $u_k \to u_0$ in $L_2(0,1)$, because the sequence $\{u_k\}$ is minimizing. Thus, each minimizing sequence tends to the optimal control. Therefore, this optimal control problem is well-posed by Tichonov. $\square$

Remark 7.26 One can prove that the inequality (7.14) that is called the **strong convexity of the functional** guarantees the well-posedness by Tikhonov of the problem for the general case too.

Now we return to the general problem of this step of analysis. There are properties of the sequentially optimal control.

Example 7.9 Consider the extended optimal control problem for Example 7.7. Suppose the control sequences $\{u_k\}$ and $\{v_k\}$ are equivalent with respect to the relation φ, if its elements are unlimitedly close with respect to the norm of $L_2(0,1)$ (see also Example 7.5). The control sequence $\{u_k\}$ with convergent correspondent sequence of functionals determines the sequential control

$$[u_k] = \left\{\{v_k\} \in U \mid v_k = u_k + w_k, \ w_k \to 0 \ \text{in} \ L_2(0,1)\right\}.$$

Determine the functional

$$J([u_k]) = \lim_{k \to \infty} I(u_k) = \lim_{k \to \infty} \|x_k\|^2$$

on the set V of sequential controls. Consider the minimization problem of the functional J on the set V. This is the extended optimal control problem. By the lower boundedness of the functional I on the set U, this problem has a unique solution. Suppose the sequential control u is sequentially optimal, and v is an arbitrary sequential control, i.e.

$$v = [v_k] = \Big\{\{v_k + w_k\} \in U \mid w_k \to 0 \text{ in } L_2(0,1)\Big\}.$$

We have the inequality

$$J[\sigma v + (1-\sigma)u] - J(u) \geq 0 \ \forall \sigma \in (0,1) \tag{7.15}$$

that is the analogue of (7.6). By the definition of the functional J, we have

$$J(u) = \lim_{k\to\infty} \|x_k\|^2, \quad J[\sigma v + (1-\sigma)u] = \lim_{k\to\infty} \|z_k\|^2.$$

The functions x_k and z_k are the states of the considered system for the controls u_k and $\sigma v_k + (1\sigma)u_k$. Then we transform the inequality (7.15) to

$$\lim_{k\to\infty} \|z_k\|^2 - \lim_{k\to\infty} \|x_k\|^2 \geq 0. \tag{7.16}$$

Determine the equality

$$z_k = \sigma y_k + (1-\sigma)x_k = x_k + \sigma(y_k - x_k)$$

that is the analogue of the formula (7.11). Then we get the equality

$$\|z_k\|^2 = \|x_k\|^2 + 2\sigma(x_k, y_k - x_k) + \sigma^2\|y_k - x_k\|^2 \tag{7.17}$$

with scalar product of the space $L_2(0,1)$.

Using the idea of Example 7.5, determine the function p_k from the system

$$\dot{p}_k(t) = -x_k(t), \ t \in (0,1); \ p(1) = 0.$$

This is the analogue of the problem (7.7). Multiply the equality

$$\dot{x}_k - \dot{y}_k = u_k - v_k$$

by the function p_k. Integrating it by the unit interval, after integration by parts we obtain

$$(x_k, y_k - x_k) = (p_k, v_k - u_k).$$

Now the equality (7.17) can be transformed to

$$\|z_k\|^2 = \|x_k\|^2 + 2\sigma(p_k, v_k - u_k) + \sigma^2\|y_k - x_k\|^2.$$

Using the inequality (7.16), we get

$$\lim_{k\to\infty} (p_k, v_k - u_k) + \frac{\sigma}{2} \lim_{k\to\infty} \|y_k - x_k\|^2 \geq 0.$$

184   *Sequential Models of Mathematical Physics*

Passing to the limit as $\sigma \to 0$, we have

$$\lim_{k \to \infty} \left(p_k, v_k - u_k\right) \geq 0.$$

By the transformation of Example 7.5, after extracting of subsequences we obtain the convergence $u_k \to u'$ weakly in $L_2(0,1)$, $v_k \to v'$ weakly in $L_2(0,1)$, and $p_k \to p'$ strongly in $L_2(0,1)$. From the last inequality it follows that

$$\left(p', v' - u'\right) \geq 0. \tag{7.18}$$

The function v' here is the weak limit of the arbitrary sequence of admissible controls. Using the arbitrariness of the function v', we prove that the inequality (7.18) is realized at least if $p' = 0$. Then we have (see Example 7.5) $u' = 0$. Thus, the sequential control is generated by the control sequence with zero weak limit. The corresponding sequence of the functionals tends to zero too. Therefore, this sequence will be minimizing, because the criterion of optimality is non-negative. Thus, we found the class of minimizing sequences for the considered problem. $\square$

Remark 7.27 One can prove that the minimizing sequence $\{u_k\}$ of Example 7.7 tends to zero weakly in $L_2(0,1)$.

Remark 7.28 It is possible that other minimizing sequences exist for the considered optimal control problem.

The sequentially optimal control for the considered insolvable optimal control problem was generated by the sequences of admissible control satisfying two conditions $u_k \to 0$ weakly in $L_2(0,1)$ and $\|u_k\| \to 1$. This sequential control is singular. Now the second condition is not applied. Therefore, the sequentially optimal control is generated by the sequences of admissible controls that tend to zero strongly. Hence, this is regular sequential control. By the way, the zero control that is the optimal for the considered example satisfies the variational inequality (7.18). This is the necessary and sufficient condition of optimality now.

Remark 7.29 The formal necessary condition of optimality for Example 7.3 is insolvable.

Thus, the sequentially optimal control for the solvable optimal control problems that is ill-posed by Tikhonov is regular. However, this is generated by convergent and divergent minimizing sequences of usual controls. Properties of the sequentially optimal controls for the different optimal control problems are described in Table 7.4.

7.8   Conclusions

1. The determination of mathematical models of physical phenomena is connected with the procedure for passing to the limit.

TABLE 7.4: Properties of sequentially optimal controls

property of optimal control problem	property of minimized functional	example of the integrand	sequentially optimal control	generating minimizing sequence
well-posedness by Tikhonov	strong convexity	$x^2 + u^2$	regular	convergent
single-valued solvability	strict convexity	x^2	regular	convergent and divergent
non-single-valued solvability	non-convexity	$-x^2 - u^2$	regular	convergent and divergent
non-solvability	non-convexity	$x^2 + u^2$	singular	divergent

2. The practical justification of the passage to the limit is based on the Cauchy criterion that is obtaining of the convergence of the fundamental sequence.

3. The Cauchy criterion is true for the complete spaces only.

4. The non-complete space can be extended to the complete one that consists of all equivalence classes of fundamental sequences of the initial space.

5. Each element of the completion can be approximated by elements of the initial space.

6. Using the completion technique that is the basis of the sequential method, one can extend the non-complete space of rational numbers to the complete spaces of real or p-adic numbers.

7. One applies the sequential objects in the optimal control theory too.

8. The optimal control problems can be insolvable.

9. The objects of the analysis for the insolvable optimal control problems are the minimizing sequences.

10. The equivalence classes of the sequences of usual controls with convergent sequences of values of the functionals are the sequential controls.

11. The extended optimal control problem is the minimization problem for the prolongation of the given functional to the set of the sequential controls.

12. The extended optimal control problem has the unique solution if the minimized functional is lower bounded on the set of admissible controls.

13. The minimum of the functional for the extended problem is equal to the least lower bound of the initial functional.

14. An element of minimizing sequence can be chosen as an approximate solution of the initial optimal control problem.

15. The sequentially optimal control for the optimal control problems with non-unique solution and the ill-posed problem by Tikhonov can be generated by convergent and divergent sequences of usual controls.

After consideration of the sequential objects of numbers theory and optimal control theory, we consider also the very important class of sequential objects. These are the distributions. This theory is very important for mathematical physics problems (see Chapter 2). Then we will come very close to our general problem that is the justification of the determination of mathematical models of physical phenomena.

Notes

[1]Optimal control theory deals with the problem of finding a control law for a given system such that a certain optimality criterion is achieved under the given constraints, see [8], [54], [140], [136], [73], [86], [97], [113], [199].

[2]The easiest necessary condition of extremum for the polynomial was obtained by *Pierre Fermat*. This result was extended to general functions by *Gottfried Wilhelm Leibniz* in 1684.

[3]The problems of calculus of variations are considered, for example, in [22], [50], [58], [78], [149].

[4]The base of the solvability theorems for extremum problems is the existence theorem of the exact lower bound of the lower bounded set proved by *Bernard Bolzano* in 1817 and the **Weierstrass theorem** of the existence of the minimum of the continuous functions on the closed interval. The first result of the solvability of the infinite dimensional extremum problem is the justification of the Dirichlet principle by *David Hilbert*. The general existence theorems for the optimal control problems are given, for example, in [49], [116], [192].

[5]Example 7.3 of the insolvable optimal control problem is considered in [199].

[6]Other insolvable optimization control problems are considered, for example, in [18], [33], [131], [181]. The first example of an insolvable problem of variation calculus was given by *Karl Theodor Wilhelm Weierstrass*

[7]The problem of analyzing unsolvable extremal problems was proposed by *David Hilbert*. The first results in this direction were by obtained by *Laurence Young* [204] and *Edward McShane* [125]; see also [33], [49], [97], [181], [199].

[8]The sequential controls were determined in [165], [163]; see also [164], [166].

[9]Different extension methods of optimal control problems are considered, for example, in [33], [49], [97], [181], [199].

[10]The different forms of the approximate solutions of the optimal control problems for nonlinear infinite dimensional systems are used in [167], [168], [170].

[11]The first problem involving a **variational inequality** was the Signorini

problem, posed by *Antonio Signorini* in 1959 and solved by *Gaetano Fichera* in 1963. *Jacques-Louis Lions* used the variational inequality as a necessary condition of optimality [116]. The theory of variational inequality is given in [46], [60], [84].

[12]The partial case of the Banach–Alaoglu theorem was considered by *Stefan Banach* in 1932, and the proof for the general case was published in 1940 by *Leonidas Alaoglu*.

[13]The general statement of the Banach–Alaoglu theorem is given, for example, in [147].

[14]The Rellich–Kondrashov theorem is considered, for example, in [1], [101], [177].

[15]Example 7.6 of non-uniqueness of the optimal control is considered in [160]. Others examples of non-uniqueness for the optimal control problems are given in [56].

[16]The example 7.7 of ill-posed optimal control problem is considered in [192]. The well-posedness of the optimal control problems and the regularization method for ill-posed problems are discussed there too.

[17]Well-posedness by Tikhonov of optimal control problems is considered in [192], [205].

[18]Well-posedness by Hadamard of optimal control problems is considered in [205]. This is an extension of the well-posedness of mathematical physics problems that is the unique solvability of the problem and the continuity of its solution with respect to the parameters of the system, see [62], [185].

Chapter 8

Distributions

We approach the conclusion of our analysis (see Figure 8.1). The determination of mathematical models of physical processes is often associated with passage to the limit in the balance relations describing a certain conservation law in an elementary volume (see Chapter 1 and Chapter 2). The justification of the passage to the limit in classical mathematical analysis is based on the Cauchy criterion, according to which any fundamental sequence of real numbers converges (see Chapter 3). However, the Cauchy criterion is applicable only in complete spaces, of which there are not too many (see Chapter 4). Nevertheless, any metric space can be extended to completion, and any of its elements can be arbitrarily closely approximated by elements of the original space (see Chapter 5). The basic idea of the sequential method is that any fundamental sequence of an arbitrary space becomes convergent in a certain sense after considering its completion.

We used the sequential method for the definition of real numbers (see Chapter 4 and Chapter 5), p-adic numbers (see Chapter 6), and sequential controls (see Chapter 7). Another extremely important sequential object is the distribution considered below. This is important to us not only as another area of application of the sequential method, but also because the theory of distributions underlies the concept of the generalized solution of problems of mathematical physics (see Chapter 2). In addition, this is directly related to the main topic of our study.

We know that mathematical objects can have different interpretations. It is true for real numbers (Chapter 4 and Chapter 5) and p-adic numbers (Chapter 6). This is true for distributions too. At first, we consider the standard definition of distributions by Schwartz. This is a linear continuous functional on a set of functions here; i.e. the element of the adjoint space to this considered function space. Therefore, we need to describe, first, the used function space (Section 8.1). Then we can give the definition of distributions and describe their general properties (Section 8.2). One of the most important properties

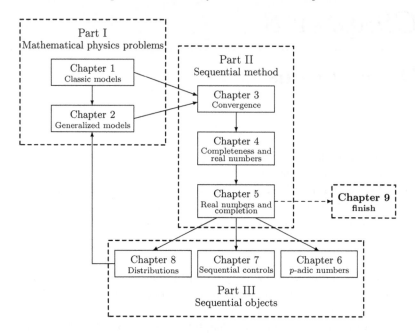

FIGURE 8.1: Structure of the book.

here is the possibility of differentiation. We give the strict definition of the generalized derivatives, which was applied in Chapter 2. We consider also the sequential interpretation of distributions theory (Section 8.3). We shall make sure that the distributions and their properties can be described by the sequential technique. Finally, the considered generalized derivatives are used for the definition of Sobolev spaces (Section 8.4), which are the most important class of functional spaces of mathematical physics.

8.1   Test functions

In the definition of the generalized solution of mathematical physics problems in Chapter 2, we used the concept of the generalized derivative. By Definition 2.4, the object du/dx is called the **generalized derivative** of the function u on the interval (a, b), if it satisfies the equality

$$\int\limits_a^b \frac{du(x)}{dx}\lambda(x)dx = -\int\limits_a^b u(x)\frac{d\lambda(x)}{dx}dx \qquad (8.1)$$

for any smooth enough function λ with zero values at the points a and b.

We did not stipulate in advance what requirements are imposed on the function λ here. However, it is clear that this is so smooth that both integrals of the equality (8.1) make sense. In Chapter 2, one considered examples that clarify the situation to some extent. Particularly, if we choose as u the continuous function v such that $v(x) = |x|$ (see Example 2.2) and the piecewise continuous function w that equals to -1 for the negative values of the argument and 1 for its positive values (generalized derivative of u, see Example 2.3), then the function λ is continuous. However, the function λ is differentiable for the application of the equality (8.1) to the generalized derivative y of w characterized by the equality (see Example 2.4)

$$\int_a^b y(x)\lambda(x)dx = 2\lambda(0).$$

Obviously, the weaker the properties of the object under consideration, the stronger the requirements must be used for the function λ, so that the terms under the integrals of the equality (8.1) become integrable.

The properties of the function worsen as a rule after differentiation. Particularly, the continuous non-smooth function v is the generalized derivative (and classic too) of the differentiable function u determined by the equality (see Example 2.1)

$$u(x) = \frac{1}{2}x|x|.$$

The generalized derivative of the continuous non-smooth function v is the discontinuous function w. Its generalized derivative is a strange object y that is not even a function. We would like to extend the differentiation procedure to the widest possible class of objects. It is desirable that this class is so wide that differentiation would not derive from this set.

Remark 8.1 We often considered the extension of a class of objects such that the retention on this set is guaranteed after performing some procedure (see Table 8.1). For example, the transition from a given metric space to its completion ensures that any fundamental sequence there is convergent. Extending the monoid to the group (see Chapter 5) ensures the invertibility of any element of the set. The transition from the set of natural (respectively, integer and rational) numbers to integers (respectively, to non-zero rational and complex) numbers guarantees the solvability of the additive (respectively, multiplicative and algebraic) equation. The extension of the set of rational numbers to the set of real Dedekind numbers guarantees that any cut of the considered set of objects will be determined by the object of the given class. In all these examples, a specific completion of the original class of objects with respect to the given procedure is realized.

It is necessary to choose the class of functions λ with very good properties for obtaining the extremely large class of objects u, where the differentiation makes sense. We give the general definitions for the multidimensional case. Let Ω be an open set of the n-dimensional Euclid space $\mathbb{R}^n$. Particularly, for the one-dimensional case the set Ω can be an interval (a, b).

TABLE 8.1:   Examples of completions

initial class of objects	procedure	extended class of objects
metric space	convergence of any fundamental sequences	completion of the metric space
monoid	invertibility of any elements	group
set of natural numbers	solvability of any additive equations	set of integer numbers
set of integer numbers	solvability of any multiplicative equations	set of non-zero rational numbers
set of rational numbers	solvability of any algebraic equations	set of complex numbers
set of rational numbers	definition of any cut by a rational number	set of Dedekind real numbers
set of functions	differentiation of any functions	set of distributions

Definition 8.1 *The **support** of the function u on the set Ω is the smallest closed subset* $\mathrm{supp}(u)$ *of Ω such that u is equal to zero outside* $\mathrm{supp}(u)$.

Remark 8.2 The smallest closed subset of the given set is called its **closure**[1].

Example 8.1 *Support of the function.* Consider the function of one variable determined by the formula (see Figure 8.2)

$$u(x) = \begin{cases} 1 - x^2, & \text{if } x < 1, \\ 0, & \text{if } x \geq 1. \end{cases}$$

This function is not equal to zero on the interval $(-1, 1)$. However, this set is not closed. The smallest closed set that includes this interval is $[-1, 1]$. Therefore, $\mathrm{supp}(u) = [-1, 1]$. $\square$

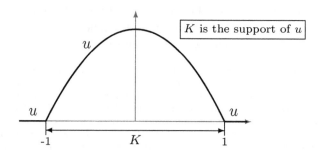

FIGURE 8.2: Support of the function.

We consider now the functions on the set Ω with **compact support**; i.e. this support is bounded[2]. Particularly, the function u from Example 8.1 has compact support.

Example 8.2 *Function with non-compact support.* Consider the function u determined by the formula (see Figure 8.3)

$$u(x) = \frac{1}{1 + x^2}.$$

This function vanishes at infinity, i.e. $u(x) \to 0$ if $|x| \to \infty$. However, its support is the set $\mathbb{R}$ that is not compact. □

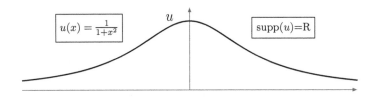

FIGURE 8.3: Function with non-compact support.

Definition 8.2 *The infinite differentiable function on the set Ω with compact support is called the* **test function**[3].

Remark 8.3 In reality, the function with compact support is called *finite*. This notion is used, as a rule, if we consider it as an independent object. However, now we introduce a set of all functions that will be testable in relation to the notion of distribution introduced below.

We shall determine the function λ from the equality (8.1) as the test function on the set $\Omega = (a, b)$. Denote by $D(\Omega)$ the set of all test functions on set Ω. Note that for any functions of the set $D(\Omega)$ there exists its derivative that is the element of the set $D(\Omega)$.

Remark 8.4 By the final result, the differentiation is the first order operation on the set $D(\Omega)$.

Determine the standard operations of addition and multiplication by numbers on the set $D(\Omega)$ by the formulas

$$(\lambda + \mu)(x) = \lambda(x) + \mu(x), \; x \in \Omega;$$

$$(c\lambda)(x) = c\lambda(x), \; x \in \Omega$$

for all $\lambda, \mu \in D(\Omega)$. Now $D(\Omega)$ is the linear space.

Suppose the sequence $\{\lambda_k\}$ of the set $D(\Omega)$ tends to zero function, if and only if there exists a compact set K of Ω such that the supports of all functions λ_k belong to K, and the sequences of the derivatives of arbitrary order of λ_k tend to zero uniformly on K. It means that $\lambda_k(x) \to 0$; besides the velocity of convergence does not depend from the point x of K; and the analogical property is true for all derivatives of λ_k. Particularly, for the one-dimensional case, we have the convergence

$$\sup_{x \in K} \left| \frac{d^m \lambda_k(x)}{dx^m} \right| \to 0, \ m = 0, 1, \dots .$$

Consider now the n-dimensional case with points $x = (x_1, \dots, x_n)$. The vector $\alpha = (\alpha_1, \dots, \alpha_n)$ with non-negative integer components is called the **multiindex**. Determine the **differentiation operator**

$$D^\alpha = \frac{\partial^{|\alpha|}}{\partial x_1^{\alpha_1} \dots \partial x_n^{\alpha_n}}$$

of order α, where $|\alpha| = \alpha_1 + \dots + \alpha_n$. The sequence $\{\lambda_k\}$ of the set $D(\Omega)$ tends to zero function, if

$$\sup_{x \in K} \left| D^\alpha \lambda_k(x) \right| \to 0$$

for all α. Now the sequence $\{\lambda_k\}$ **tends** to a function λ in $D(\Omega)$, if the sequence of the difference $\{\lambda_k - \lambda\}$ tends to zero. The set $D(\Omega)$ with this form of convergence is a topological space.

Remark 8.5 This space is not metrisable; that is this convergence is not described by any metric.

Suppose we have the convergence of the sequences of the infinite differentiable functions $\lambda_k \to \lambda$ and $\mu_k \to \mu$. Then we have $(\lambda_k + \mu_k) \to (\lambda + \mu)$ and $a\lambda_k \to a\lambda$ for any number a. Therefore, the operations of addition and multiplication by the number are continuous. Therefore, the set $D(\Omega)$ with considered operation and convergence is the linear topological space. It very important that this space is complete[4].

Example 8.3 *Non-complete metrisable space*[5]. Consider the set $D(\Omega)$ with determined operations and easier topology that are relevant to the following convergence. Suppose the sequence $\{\lambda_k\}$ of the set $D(\Omega)$ tends to the function λ, if $D^\alpha \lambda_k \to D^\alpha \lambda$ uniformly on Ω for all multiindex α. Then we obtain the linear topological space with metrisable topology. Consider the one-dimensional case for $\Omega = \mathbb{R}$. Let φ be a function from $D(\Omega)$ with support $[0,1]$ that is positive on the interval $(0,1)$ (see Figure 8.4).

Determine the sequence of functions $\{\lambda_k\}$ by the equalities (see Figure 8.5)

$$\lambda_k(x) = \sum_{i=0}^{k-1} \frac{\varphi(x - i)}{i + 1}.$$

The function λ_k is infinite differentiable; besides, its derivatives are determined by the formula

$$\frac{d^m \lambda_k(x)}{dx^m} = \sum_{i=0}^{k-1} \frac{\varphi^{(m)}(x - i)}{i + 1}.$$

The support of the function λ_k is the interval $[0, k]$ (see Figure 8.5). For any point x we have the convergence $\lambda_k(x) \to \lambda(x)$, where

$$\lambda(x) = \sum_{i=0}^{\infty} \frac{\varphi(x-i)}{i+1}.$$

We have the equality

$$\lambda_{k+p}(x) - \lambda_k(x) = \sum_{i=k}^{k+p-1} \frac{\varphi(x-i)}{i+1}.$$

Therefore, we get

$$\sup_{x\in\mathbb{R}} \left|\lambda_{k+p}(x) - \lambda_k(x)\right| = \frac{M}{k+1},$$

where $M = \max \varphi$. Then

$$\lim_{k\to\infty} \sup_{x\in\mathbb{R}} \left|\lambda_{k+p}(x) - \lambda_k(x)\right| = 0.$$

The relation

$$\lim_{k\to\infty} \sup_{x\in\mathbb{R}} \left|\frac{d^m\lambda_{k+p}(x)}{dx^m} - \frac{d^m\lambda_k(x)}{dx^m}\right| = 0, \; m = 1, 2, \dots$$

can be determined analogically. Thus, the sequence $\{\lambda_k\}$ is fundamental. However, this is not convergent with respect to the considered topology of $D(\Omega)$. Particularly, the function λ is infinite differentiable, but its support is the non-compact set of non-negative numbers. Hence, the space $D(\Omega)$ is non-complete with respect to the given form of convergence. This result clarifies the choice not of this but of the previously introduced non-metrisable topology of the space of test functions. $\square$

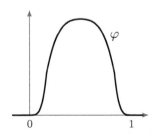

FIGURE 8.4: Function φ of Example 8.3.

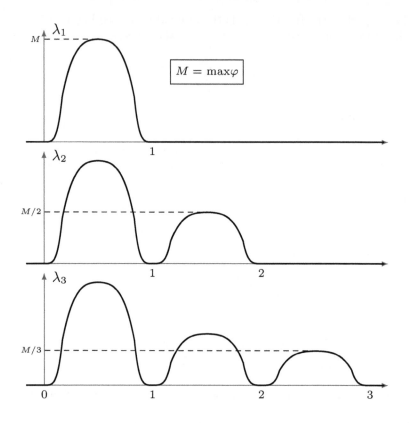

FIGURE 8.5: Sequence $\{\lambda_k\}$.

Remark 8.6 We shall determine soon another notion of the fundamentality (see Definition 8.6). The set $D(\Omega)$ is non-complete too with respect to that form of fundamentality. It will be the basis of the definition of the distribution as a sequential object.

Now we determine the distributions on the basis of the linear topological space of test functions.

8.2   Schwartz distributions

We know (see Chapter 3) that for any linear topological space X one determines its adjoint space that is the set X' of all continuous functionals on the set X.

Definition 8.3 *The **distribution** (more exactly, the **Schwartz distribution**) on the set Ω is an element of the adjoint space $D'(\Omega)$ to the space $D(\Omega)^6$.*

Remark 8.7 The distributions can have a real physical sense. They can describe, for example, the density of the material point, the intensity of the instantaneous source, the point charge, etc. In principle, any external influences on the system acting at individual points of the given region are described by distributions.

Denote the value of the linear continuous functional u at the point λ of the space $D(\Omega)$ by $\langle u, \lambda \rangle$. Note that the distributions do not have any values at the concrete points; there are no usual functions.

Remark 8.8 Therefore, one sometimes uses the nomination ***generalized function*** instead of distribution.

Remark 8.9 Sometimes we use the denotation $u(x)$ for the distribution u, because the distribution can be reduced to usual functions at concrete points.

Example 8.4 ***Local integrable function.*** Consider a local integrable function φ on the set Ω; i.e. this is integrable on the arbitrary compact subset of Ω. This determines the object u_φ that satisfies the equality

$$\langle u_\varphi, \lambda \rangle = \int_\Omega \varphi(x)\lambda(x)dx \ \forall \lambda \in D(\Omega). \tag{8.2}$$

Indeed, for any function λ we have the number at the right-hand side of this equality. Therefore, we have the functional. It is linear because of the linearity of the integration. If we have the converged sequence of the infinite differentiable functions, then we obtain the convergence of the relevant sequence of the integrals. Hence, the functional u_φ is continuous. Thus, it satisfies Definition 8.1. Therefore, this is the distribution, in reality.

Suppose φ_1 and φ_1 are local integrable functions; besides the following equality holds

$$\langle u_{\varphi_1}, \lambda \rangle = \langle u_{\varphi_2}, \lambda \rangle \ \forall \lambda \in D(\Omega).$$

Then $\varphi_1(x) = \varphi_2(x)$ for almost any point x, i.e. for all points except the points of a set with zero measure[7]. Therefore, the operator $\varphi \to \langle u_\varphi, \cdot \rangle$, which maps any local function to the relevant distribution determined by the equality (8.2), gives embedding of the space of local integrable functions to the space distributions.□

Remark 8.10 We can identify the local integrable function and the distribution, which is determined by it.

Definition 8.4 *The distribution is **regular** if it can be determined by a local integrable function. The distribution is **singular** if this is not regular.*

Example 8.5 *Jump function.* Determine the regular distribution u_φ by the equality

$$\langle u_\varphi, \lambda \rangle = \int\limits_0^\infty \lambda(x)dx \ \forall \lambda \in D(\Omega).$$

The relevant function φ is

$$\varphi(x) = \begin{cases} 0, & \text{if } x < 0, \\ 1, & \text{if } x \geq 0. \end{cases}$$

This distribution is called the jump function. □

Example 8.6 *δ-function.* For a point $\xi \in \Omega$ determine the object δ_ξ by the equality

$$\langle \delta_\xi, \lambda \rangle = \lambda(\xi) \ \forall \lambda \in D(\Omega).$$

Obviously, this is the linear continuous functional on the space $D(\Omega)$. Therefore, this is the distribution. It is called the δ-function at the point ξ. We could write formally

$$\langle \delta_\xi, \lambda \rangle = \int\limits_\Omega \delta_\xi(x)\lambda(x)dx \ \forall \lambda \in D(\Omega).$$

However, there does not exist any local integrable function that determines this distribution. This is not associated with any function. □

The δ-function is the standard example of the singular distribution. By $\delta(x - \xi)$ one describes the δ-function at the point ξ.

Remark 8.11 The δ-function describes, for example, the density of the unit mass at the given point.

Remark 8.12 The δ-function can be described also by measure theory[8].

Determine the standard algebraic operations on the set of distributions by the equalities

$$\langle u + v, \lambda \rangle = \langle u, \lambda \rangle + \langle v, \lambda \rangle \ \forall \lambda \in D(\Omega),$$

$$\langle au, \lambda \rangle = a\langle u, \lambda \rangle \ \forall \lambda \in D(\Omega)$$

for all distributions u, v and number a. Then the set $D'(\Omega)$ is the linear space.

Now determine a topology by the following convergence. The sequence of distributions $\{u_k\}$ *tends* to a distribution u, if the numerical sequence $\{\langle u_k, \lambda \rangle\}$ tends to the number $\langle u, \lambda \rangle$ for all $\lambda \in D(\Omega)$.

Remark 8.13 The relevant topological space is not metrisable.

Obviously, the linear operations are continuous. Therefore, the set of distributions with given operations and convergence is the linear topological space.

Example 8.7 *Schwartz example*[9]. Obviously, we can not only add, but also multiply the infinite differentiable functions. However, the definition of the multiplication for the distributions is not clear. At first, the definition of the distribution is based on the integration. However, the product of integrable functions can be non-integrable. The integration operation is by its nature additive, but not multiplicative, which creates serious obstacles to the interpretation of the product of distributions as a linear continuous functional.

Consider now a one-dimensional set Ω, which included zero. Define three distributions u, v and w by the equalities

$$u(x) = 1/x, \ v(x) = x, \ w(x) = \delta(x). \tag{8.3}$$

However, there is not exist any local integrable function that is determined this distribution. This is not associated with any function.

The object w here is δ-function at zero, i.e. the singular distribution, and v is the usual function. The object u has the singularity at zero point. Therefore, this is the distribution that is equal to the usual function $1/x$ outside an arbitrary neighborhood of zero[10].

Suppose one determine a multiplication on the set $D'(\Omega)$. Apparently, multiplying the objects u and v should yield the function that is identically equal to one. Then we have the equality

$$(u \cdot v) \cdot w = 1 \cdot w = \delta.$$

The product $v \cdot w$ satisfies the equalities

$$\langle v \cdot w, \lambda \rangle = \int_{\Omega} x\delta(x)\lambda(x)dx = \int_{\Omega} \left[x\lambda(x) \right]\delta(x)dx = 0 \cdot \lambda(0) = 0 \ \forall \lambda \in D(\Omega)$$

by the definition of δ-function. Therefore, we get $v \cdot w = 0$, i.e. this product is zero element of the space $D'(\Omega)$. Now we have

$$u \cdot (v \cdot w) = 0.$$

Thus, the following inequality holds

$$(u \cdot v) \cdot w \neq u \cdot (v \cdot w).$$

Hence, the multiplication of distributions is non-associative operation. This result shows that there are extremely serious difficulties in determining a natural operation of multiplication on the set of distributions. The naturalness of the operation here means that in the case when we operate with regular distributions that are, in fact, usual functions, the result will be a regular distribution corresponding to the natural product of these usual functions[11].□

Remark 8.14 We might be able to reconcile ourselves to the absence of associativity of distributions multiplication, citing the fact that sometimes we consider the algebraic objects with non-commutative multiplication[12]. However, the absence of the associativity is more serious trouble than the absence of the commutativity[13]. We shall consider, particularly, a very strange corollary of the Schwartz example (see Remark 8.19).

Remark 8.15 The multiplication of distribution is very useful, for example, for nonlinear differential equations. Sometimes one consider equations that contain the degree of unknown functions, the product of two different unknown functions, the product of unknown functions by its derivatives, etc. If we interpret the solution of the problem as a distribution, then we have the necessity to consider the multiplication of distribution.

The very important property of the distributions theory is the possibility of the determination of the derivatives for the arbitrary distribution[14].

Definition 8.5 *The **generalized** α order **derivative** of the distribution u is the distribution $D^\alpha u$, determined by the equality*

$$\langle D^\alpha u, \lambda \rangle = (-1)^{|\alpha|} \langle u, D^\alpha \lambda \rangle \ \forall \lambda \in D(\Omega). \tag{8.4}$$

Prove that the generalized derivative $D^\alpha u$, which is determined by the equality (8.4), is, in reality, the distribution. Indeed, we have the classic operator of differentiation at the right-hand side of the equality (8.3). Therefore, $D^\alpha \lambda$ is the element of the space $D(\Omega)$ for all $\lambda \in D^\alpha u$. Then for all infinite differentiable function λ we have the number $\langle D^\alpha u, \lambda \rangle$. Therefore, $D^\alpha u$ is the functional on the space $D(\Omega)$. Besides, the map $\lambda \to \langle D^\alpha u, \lambda \rangle$ is linear, because of the linearity of the classic operator of differentiation and the map $\mu \to \langle v, \mu \rangle$. Finally, if $\lambda_k \to \lambda$ in $D(\Omega)$, then we have $D^\alpha \lambda_k \to D^\alpha \lambda$ in $D(\Omega)$ by the definition of the convergence of the space $D(\Omega)$. Hence, $\langle u, D^\alpha \lambda_k \rangle \to \langle u, D^\alpha \lambda \rangle$, because u is the continuous functional on $D(\Omega)$. Thus, $\langle D^\alpha u, \lambda_k \rangle \to \langle D^\alpha u, \lambda \rangle$, i.e. $D^\alpha u$ is the linear continuous functional on $D(\Omega)$. Therefore, this is the distribution by Definition 8.3.

Remark 8.16 Thus, the generalized differentiation is the first order operation on the set of distributions.

One uses also the integral form of the equality (8.4)

$$\int_\Omega D^\alpha u \lambda dx = (-1)^{|\alpha|} \int_\Omega u D^\alpha \lambda dx \ \forall \lambda \in D(\Omega).$$

We have the following formula for one-dimensional case

$$\int_a^b \frac{d^m u}{dx^m} \lambda dx = (-1)^m \int_a^b \frac{d^m \lambda}{dx^m} u dx \ \forall \lambda \in D(a,b).$$

This is agreed with our previous definition of the generalized derivative (see Chapter 2), particularly, with formula (8.1).

Remark 8.17 We know that the properties of a function worsen after differentiation. Particularly, after differentiation of the function of $C^m(\overline{\Omega})$ (this is the set of m time continuously differentiable functions on the closure of the set $D(\Omega)$) we obtain the function of the space $C^{m-1}(\overline{\Omega})$ (see Figure 8.6). Thus, one decreases the order of differentiability. Therefore, the degree of smoothness of the function worsens. The analogical result is true for the Sobolev spaces $W_p^m(\Omega)$ (see Section 8.4). However, there exist two classes of objects, where this result is not realized. This is the space $D(\Omega)$ of infinite differentiable functions. There are so good objects that are remain infinite differentiable after differentiation. This is also the space $D'(\Omega)$ of distributions. There are so bad objects that it cannot to become worse after differentiation.

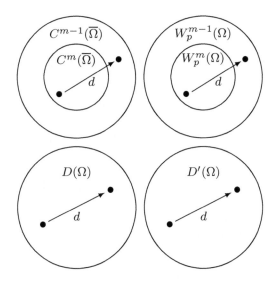

FIGURE 8.6: Action of the operator d of generalized differentiation.

8.3 Sequential distributions

We have already known that mathematical notions can have different interpretation. This is true for the distributions too. Particularly, there exists its sequential interpretation.

One of the general result of the Cantors theory of the real numbers is the possibility of determining of the irrational real numbers as limits of sequences of rational numbers. Try to use this idea for the distributions.

Example 8.8 δ-*function*. Consider the singular distribution $\delta = \delta(x)$. Determine the functional sequence $\{u_k\}$ by the equality (see Figure 8.7)

$$u_k(x) = \begin{cases} k/2, & \text{if } |x| \leq 1/k, \\ 0, & \text{if } |x| > 1/k, \end{cases}$$

where $k = 1, 2, \dots$. For all function λ of the space $D(\mathbb{R})$ we have the equality

$$\int_{-\infty}^{\infty} \lambda(x)u_k(x)dx = \frac{k}{2}\int_{-1/k}^{1/k} \lambda(x)dx = \lambda(\xi_k),$$

where $|\xi_k| \leq 1/k$ by Mean value theorem (see Theorem 2.3). Pass to the limit as $k \to \infty$. We have

$$\lim_{k \to \infty} \int_{-\infty}^{\infty} \lambda(x) u_k(x) dx = \lim_{k \to \infty} \left[\frac{k}{2} \int_{-1/k}^{1/k} \lambda(x) dx \right] = \lambda(0) = \int_{-\infty}^{\infty} \lambda(x) \delta(x) dx,$$

because the function λ is continuous. Therefore, we have the convergence

$$\int_{-\infty}^{\infty} \lambda(x) u_k(x) dx \to \int_{-\infty}^{\infty} \lambda(x) \delta(x) dx.$$

Thus, we can interpreted δ-function as the limit of the sequence of classic functions $\{u_k\}$.

Moreover, we can obtain the analogical result with using the sequence of the infinite differential functions

$$u_k(x) = \frac{k}{\sqrt{2\pi}} e^{-n^2 x^2/2}, \ k = 1, 2, \dots .$$

Thus, δ-function can be obtained as the limit of the sequence of infinite differentiable functions too. $\square$

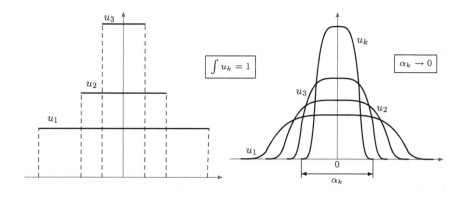

FIGURE 8.7: Approximating sequences for δ-function.

Thus, the singular distribution δ can be approximated by infinite differentiable functions similar the irrational number π can be approximated by rational numbers. Moreover, we have the following result[15].

Theorem 8.1 *For any $u \in D'(\Omega)$ there exists a sequence $\{u_k\}$ of $D(\Omega)$ such that $u_k \to u$ in $D'(\Omega)$.*

Remark 8.18 By this result, the space of infinite differentiable functions $D(\Omega)$ is dense in the space of distributions similar the set of rational numbers is dense in the set of real numbers.

Remark 8.19 Return to the Schwartz example and the problem of distribution multiplication. Suppose one determine a continuous multiplication there. Therefore, from the convergence of two distribution sequences it follows the convergence of the sequence of its products. By the density of embedding of the space $D(\Omega)$ to $D'(\Omega)$, each distribution can be determined by the limit of the sequence of infinite differentiable functions. Consider now sequences $\{u_k\}$, $\{v_k\}$, and $\{w_k\}$ of $D(\Omega)$ that converges to the distributions u, v, and w, see the equalities (8.3). By the supposition of continuity of the distribution multiplication we have[16]

$$\lim_{k\to\infty}\left[(u_k v_k)w_k\right] = (u\cdot v)\cdot w = \delta,$$

$$\lim_{k\to\infty}\left[u_k(v_k w_k)\right] = u\cdot(v\cdot w) = 0.$$

Using the associativity of the usual functions, we determine that the sequence $\{u_k v_k w_k\}$ has two limits that are δ and 0. It means if the considered topological space is not Hausdorf space[17]. Non-uniqueness of the limit of the sequence is the catastrophic situation. So, not only is it impossible to introduce an associative multiplication operation on the set of distributions, consistent with the natural multiplication of usual functions. It turns out that we cannot define multiplication and a Hausdorf topology in such a way that this operation is continuous. This gives a very bad chance for the practical application of this distribution theory with continuous multiplication.

The obvious analogy between the rational and real numbers and the infinitely differentiable functions and distributions allows us to hope for the possibility of a sequential definition of distributions. Let again Ω be an open set of n-dimensional Euclid space $\mathbb{R}^n$, and $D(\Omega)$ be our space of test functions. We know that this is not metrisable. Therefore, the direct application of the completion theorem is not possible here. However, the space $D(\Omega)$ has a good enough property that is not typical for the general topological space. Indeed, we can estimate the closeness of its two arbitrary elements. The space with this property is called uniform[18]. We can determine the fundamental sequences here.

Definition 8.6 *The sequence $\{u_k\}$ of $D(\Omega)$ is called **fundamental**, if for all compact set K of Ω there exist a multiindex α and functions $\{\lambda_k\}$ of $D(\Omega)$ such that $D^\alpha \lambda_k = u_k$ on the set K, and the sequence $\{\lambda_k\}$ converges uniformly on K.*

Remark 8.20 This definition does not use the properties of uniform spaces. However, we cannot any possibility avoid these properties for the determination of the equivalence of these fundamental sequences. This is the analogue of the sequential controls definition, see Chapter 7.

By this definition, the numerical sequence $\{\lambda_k(x)\}$ has a limit, besides the velocity of the convergence does not depend from the point x of K. If the sequence $\{u_k\}$ is fundamental, then the sequence of its derivatives $\{D^\beta u_k\}$ is fundamental too for all order β. Indeed, we can choose the order $\gamma = \alpha + \beta$ with equality $D^\gamma \lambda_k = D^\beta u_k$ on and uniform convergence of the sequence $\{\lambda_k\}$ on the set K.

Remark 8.21 Note that the forms of sequence fundamentality can be diverse enough (see Table 8.2).

TABLE 8.2: Fundamentality and equivalence for different spaces

chapter	space	fundamentality of a sequence $\{u_k\}$	equivalence of sequence $\{u_k\}$	completion
4	set of rational numbers	$\lim |u_{k+p} - u_k| = 0$	$[|u_k - v_k|] \to 0$	set of real numbers
5	metric space	$\rho(u_{k+p}, u_k) = 0$	$\rho(u_k, v_k) \to 0$	completion of the metric space
7	set of control	$\exists \lim I(u_k)$	$[I(u_k) - I(v_k)] \to 0$	set of sequential control
8	set of infinite differential functions	$\exists \alpha, \exists \{\lambda_k\} \in D(\Omega):$ $u_k = D^\alpha \lambda_k,\ \exists \lim \lambda_k$	$\exists \alpha, \exists \{\lambda_k\}, \{\mu_k\} \in D(\Omega):$ $u_k = D^\alpha \lambda_k,\ v_k = D^\alpha \mu_k,$ $(\lambda_k - \mu_k) \to 0$	set of distributions

Consider the set F of all fundamental sequences on the set $D(\Omega)$. Determine the relation φ on the set F such that the condition $\{u_k\}\varphi\{v_k\}$ is true, if for all compact set K from Ω there exist an order α and sequences $\{\lambda_k\}$ and $\{\mu_k\}$ of $D(\Omega)$ such that $D^\alpha\lambda_k = u_k$, $D^\alpha\mu_k = v_k$ on K, and the sequence of differences $\{(\lambda_k - \mu_k)\}$ tends to zero uniformly on K. This relation is the equivalence on the set $D(\Omega)$.

Remark 8.22 After definition of the sequence fundamentality, the definition of its equivalence is obvious enough. Two fundamental sequence are equivalence, if their elements approach each other in the corresponding sense (see Table 8.2).

Now we give the sequential definitions of the distributions[19].

Definition 8.7 *The **sequential distributions** on the set Ω are the elements of the factor-set $S(\Omega) = F/\varphi$.*

The sequential distribution is an equivalence class of fundamental sequences as Cantor real number, p-adic number, sequential control, and point of completion of a metric space.

Determine the relation between infinite differential functions and sequential distributions. Let $\{u_k\}$ be a sequence of the space $D(\Omega)$ with infinite differentiable limit u. Therefore, there exists a compact set K of Ω such that the support of all functions u_k belong to K, and the sequences of all derivatives of u_k converge to the derivatives of u uniformly on K. Then we can choose the sequence $\{u_k\}$ as $\{\lambda_k\}$ with order $\alpha = 0$. It is obviously, that this sequence is fundamental. Therefore, it determine a distribution u' because of Definition 8.7. Another sequence of $D(\Omega)$ with limit u is equivalent to $\{u_k\}$. It determine the same distribution u'. Hence, for all infinite differentiable function u there exist a set of the fundamental sequences with limit u. Note that the stationary sequence with element u belongs to this set too. Determine the operator $A : D(\Omega) \to S(\Omega)$ such that $u = u'$ (see Figure 8.8).

Each element u' of the image of the set $D(\Omega)$ by the operator A is determine by the convergent to a same limit sequences of infinite differentiable functions. Then we can choose this limit as the pre-image of u'. Therefore, there exists the bijection between the set $D(\Omega)$ of infinite differentiable functions and the set of convergent fundamental sequences on $D(\Omega)$ (see Figure 8.8). Thus, each infinite differentiable function can be identified with a sequential distribution. This is an analogue of the Cantor real number and other sequential objects.

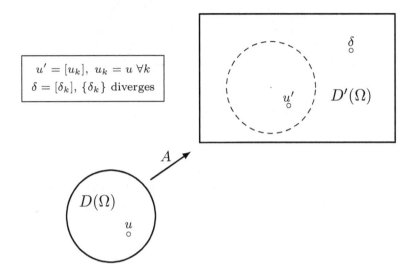

FIGURE 8.8: The set $D(\Omega)$ is isomorphic to the subset of $S(\Omega)$.

Example 8.9 *Divergent fundamental sequence of test functions* Consider one-dimensional case with $\Omega = (-1, 1)$. Determine the sequence $\{\lambda_k\}$ of $D(\Omega)$ that is tends uniformly to the function (see Figure 8.9)

$$\lambda(x) = \frac{1 - |x|}{2}.$$

The first generalized derivative of λ is equal to $1/2$ for all negative values of x, and this is equal to $-1/2$ for all its positive values. Then the second generalized derivative of λ is equal to $-\delta$. Consider now the sequence $\{u_k\}$ of $D(\Omega)$, where u_k is the second derivative of λ_k. Then this sequence is fundamental on the space $D(\Omega)$, because this is the sequence of second derivatives of the uniformly convergent functions. However, it does not have the infinite differential limit, because $-\delta$ is the distribution (see Example 2.1). Thus, the space $D(\Omega)$ is non-complete with respect to the considered notion of fundamentality. Therefore, there exists elements of the space $S(\Omega)$ that are not belong to the image of the set $D(\Omega)$ by the operator A. There are the analogues of the irrational numbers, the elements of completion of non-complete metric that are not isometric to elements of the given metric space, p-adic numbers that are not rational, and the sequential controls that are not associated with usual controls. □

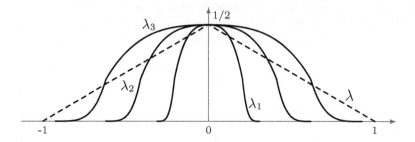

FIGURE 8.9: The sequence $\{\lambda_k\}$.

We know that each result of the theory of Cantor real numbers has an analogue of the corresponding Weierstrass and Dedekind theory. Analogically, each property of the sequential p-adic numbers has an analogue in its algebraic theory. A similar situation is true for distributions too. Determine the properties of the set of sequential distributions. We shall use here the method, which was be used for the analysis of other sequential objects. We will transfer some properties of infinitely differentiable functions to distributions, just as the basic characteristics of rational numbers were transferred to real numbers.

For all elements u and v of the set $S(\Omega)$ we choose the sequences representing them $\{u_k\}$ and $\{v_k\}$ of the space $D(\Omega)$. By its fundamentality, for any compact set K of Ω there exist multiindexes α and β, and uniformly convergent on K sequences such that the u_k and v_k are equal to the corresponding derivatives of the elements of these sequences. Earlier it was noted that the order of the derivatives in determining the fundamental sequences can be arbitrarily increased. Determine the multiindex γ such that each its component is equal to the maximal of the corresponding components of the vectors α and β. Therefore, there exists sequences $\{\lambda_k\}$ and $\{\mu_k\}$ of $D(\Omega)$ such that the following equalities hold

$$D^\gamma \lambda_k = u_k, \quad D^\gamma \mu_k = v_k.$$

Then we get

$$D^\gamma(\lambda_k + \mu_k) = (u_k + v_k),$$

besides the sequence $\{\lambda_k + \mu_k\}$ is uniformly convergent on . Thus, the sequence of sum $\{u_k + v_k\}$ is fundamental on $D(\Omega)$. By Definition 8.7, this determines a distribution that is called the sum $u + v$ of the considered distribution. One can prove that this value does not depend from the choice of the representations of the summands.

Remark 8.23 This result is substantiated in the same way as the analogous property for real numbers (see Chapter 4).

TABLE 8.3: Schwartz and sequential distributions

object	Schwartz distribution u	sequential distribution u
representation	$\langle u, \bullet \rangle$	$\lfloor u_k \rfloor$
addition	$\langle u + v, \bullet \rangle = \langle u, \bullet \rangle + \langle v, \bullet \rangle$	$\lfloor u_k \rfloor + \lfloor v_k \rfloor = \lfloor u_k + v_k \rfloor$
differentiation	$D\langle u, \bullet \rangle = -\langle u, D\bullet \rangle$	$D\lfloor u_k \rfloor = \lfloor Du_k \rfloor$

Analogically, the ***product of the distribution*** u ***by a number*** a is the distribution that is determined by the sequence $\{u_k\}$, where $\{u_k\}$ is a representation of the distribution u. Finally, the α order ***sequential derivative*** of the distribution[20]. u is the distribution that is determined by the sequence $\{D^\alpha u_k\}$. Note that the operation of addition, multiplication by number and differentiation of the sequential distributions have the same properties as the analogical operations of Schwartz distributions (see Table 8.3).

Determine the convergence on the set $S(\Omega)$. Consider a sequence $\{u_n\}$ of distributions. Let $\{u_n^k\}$ be a fundamental sequence of infinite differential functions that determines the distribution u_n. The sequence $\{u_n\}$ converges to the zero element of the linear space $S(\Omega)$, if there exists a compact subset of Ω such that the supports of all functions u_n^k belong to K, and for all n $u_n^k \to 0$ uniformly on K as $k \to \infty$. The ***convergence*** $u_n \to u$ in $S(\Omega)$ is true whenever the sequence of difference $\{u_n - u\}$ tends to zero. The set $S(\Omega)$ with corresponding topology and the determined operations is the linear topological space, i.e. the algebraic operations are continuous.

Remark 8.24 This topology is not metrisable, which, however, is not a serious obstacle to subsequent analysis.

Thus, the sequential distributions save the most important properties of the Schwartz distributions. The existence of the natural relations between two interpretations of distributions will make it possible in the future to establish a connection between the different forms of solutions of the problems of mathematical physics (see Chapter 9).

Note an explicit analogy with the theory of real numbers, where there are also the different equivalent interpretations. Although the class of equivalent fundamental sequences of rational numbers is not an infinite decimal fraction or a cut of the set of rational numbers, any action on Cantor real numbers corresponds to an absolutely analogous action on Weierstrass real numbers and Dedekind real numbers. Since mathematics does not operate with concrete objects, but with their properties, two isomorphic (that is, possessing the same properties) objects do not differ here. In this connection, certain interpretations of the distributions can be identified like the real numbers of Cantor, Weierstrass and Dedekind. As a consequence, the standard notation $D'(\Omega)$ is also used for the set of all sequential distributions.

Note, however, certain advantages of the sequential approach in the theory of distributions. This approach is preferable for the practical work with distributions. If, for example, it is necessary to implement the δ-function on the computer, then hardly anything can give the possibility of its representation in the form of a linear continuous functional on the set of infinitely differentiable functions. At the same time, its interpretation as the limit of a sequence of infinitely differentiable functions allows us to approximate this object with any degree of exactness by smooth functions (see Figure 8.7) that allow direct calculation. This circumstance to a large extent predetermines the constructivity of the sequential models of physical processes that will be determined later (see Chapter 9). Similarly, for practical work with irrational numbers like π or $\sqrt{2}$, Cantor's interpretation is preferable, which allows us to approximate them arbitrarily by the exactness of rational numbers.

Remark 8.25 The sequential method also provides additional possibilities for the definition of multiplication of distributions. In particular, let us consider two distributions u and v. By Definition 8.7, these are represented by equivalence classes of fundamental sequences of infinitely differentiable functions. Consider its arbitrary representations $\{u_k\}$ and $\{v_k\}$. If the sequence of products $\{u_k v_k\}$ is also fundamental, and the corresponding equivalence class does not depend on the choice of representing sequences, then it defines a distribution that is the product of $u \cdot v$ of the distributions u and v. Unfortunately, the product of distributions not always exists. This is true, for example, for the square of δ-function. The product's sequence of elements of fundamental sequences of infinite dimensional functions is not always fundamental. Nevertheless, while the Schwartz method, in principle, does not provide a reasonable method of distribution multiplication, in this case this possibility still persists. In this case, the method for determining multiplication is practically the same as the method for specifying other operations. Moreover, even in the general case the possibility of a reasonable interpretation of the equivalence class of the sequence of infinitely differentiable functions $\{u_k v_k\}$ is preserved as a product of the distributions of u and v, although the object obtained in this case belongs to a larger space than $D'(\Omega)$[21]. These considerations serve as additional confirmation of the effectiveness of the sequential approach in the theory of distributions.

We need only give a definition of Sobolev spaces, which are the most important class of function spaces used in mathematical physics and connected with the theory of distributions.

8.4 Sobolev spaces

The definition of Sobolev spaces is based, at first, on the spaces of integrable functions, and, second, on the notion of generalized derivatives. It is not important here, which interpretation of distributions is used for definition of the generalized derivatives. We give now some information about the spaces of integrable functions[22].

Let Ω be again an open set of the n-dimensional Euclid space. Consider measurable functions $u = u(x)$ that are Lebesgue integrable with degree $p \geq 1$, i.e. satisfying the inequality

$$\int_{\Omega} |u(x)|^p dx < \infty, \tag{8.5}$$

Remark 8.26 We do not determine the Lebesgue measure and the corresponding notions of measurable function and integral. This theory is very well described in the literature[23].

Two measurable functions on the set Ω are equivalent, if they are equal almost everywhere; i.e. with the measure of the set, where they differ is equal to zero[24]. The space $L_p(\Omega)$ is the set of the equivalent classes of measurable functions that satisfy the inequality (8.5). Determine here the operations of addition and multiplication by a number using the standard technique of definition of procedures on the factor-set. It will be addition and multiplication by the number almost at the point of Ω. Then $L_p(\Omega)$ becomes the linear space. Determine here the norm by the equality

$$\|u\| = \left[\int_\Omega |u(x)|^p dx \right]^{1/p}.$$

Then we have the linear normalized space that is Banach. The most important is the space $L_2(\Omega)$. This is the Hilbert space with scalar product

$$(u, v) = \int_\Omega u(x)v(x)dx \quad \forall u, v \in L_2(\Omega).$$

Remark 8.27 The necessity of considering not simply measurable functions, but of the corresponding equivalence classes, is due to the following circumstance. One of the general properties of the norm is its equality to zero for the zero element of the space only. However, the Lebesgue integral is equal to zero whenever the corresponding integrand is zero almost everywhere. Thus, there exists an infinite set of measurable functions, the integral of which is equal to zero[25]. All of them are equivalent to each other in the sense described above. Hence, identifying them all, i.e. passing to the corresponding equivalence class, we eliminate the considered non-uniqueness of the functions with zero integral, and hence we achieve the corresponding axiom of the norm.

Remark 8.28 We again encounter the situation, where an equivalence class is the object of consideration. However, when we consider integrable functions and elements of Sobolev spaces, we shall simply speak of functions everywhere, although in fact we are dealing with the corresponding equivalence classes.

Remark 8.29 The space $L_1(\Omega)$ is the completion of the space of continuous functions with integral norm and the space of Riemann integrable functions (see Chapter 4).

Now we can give the strict definition of Sobolev spaces[26].

Definition 8.8 *The **Sobolev space** $W_p^m(\Omega)$ is the set of all functions of the space $L_p(\Omega)$ that have the generalized derivatives of all degrees α such that $\alpha \leq m$ from the same space, where $p \geq m$, m is a natural number.*

Remark 8.30 The elements of Sobolev spaces are integrable functions. Therefore, the classical derivatives are non-applicable for this case. Hence, we use generalized derivatives that are determined by distributions theory.

The space $W_p^m(\Omega)$ is Banach with norm

$$\|u\| = \left[\int_\Omega \sum_{|\alpha| \leq m} |D^\alpha u(x)|^p dx \right]^{1/p}.$$

Definition 8.9 The **Sobolev space** $\dot{W}_p^m(\Omega)$ is the subspace of $W_p^m(\Omega)$ such that its elements are equal to zero on the boundary of Ω with all its generalized derivatives with respect to the interior normal of degree less than m.

The space $\dot{W}_p^m(\Omega)$ is Banach with the norm

$$\|u\| = \left[\int_\Omega \sum_{|\alpha|=m} |D^\alpha u(x)|^p dx\right]^{1/p}.$$

Remark 8.31 We can consider here the high derivatives only. This is impossible for the general spaces $W_p^m(\Omega)$, because for this case from equality to zero of this norm it follows that the $m-1$ degree derivatives of the considered function are constant which contradicts the property of the norm. However, for the space $\dot{W}_p^m(\Omega)$ these derivatives are equal to zero on the boundary, and we do not have any contradictions with definition of the norm.

Consider the most important case $p = 2$. Denote by $H^m(\Omega)$ and $H_0^m(\Omega)$ Sobolev spaces $W_2^m(\Omega)$ and $\dot{W}_2^m(\Omega)$. The space $H^m(\Omega)$ is Hilbert with the scalar product

$$(u, v) = \int_\Omega \sum_{|\alpha| \le m} D^\alpha u(x) D^\alpha v(x) dx.$$

The set $H_0^m(\Omega)$ is Hilbert space too; its scalar product can be determined by the easier formula

$$(u, v) = \int_\Omega \sum_{|\alpha|=m} D^\alpha u(x) D^\alpha v(x) dx.$$

The adjoint space for $H_0^m(\Omega)$ is denoted by $H^{-m}(\Omega)$. This is Hilbert space too. Note the integral formula

$$\langle u, v \rangle = \int_\Omega u(x) v(x) dx \quad \forall u \in H^{-m}(\Omega), \ v \in H_0^m(\Omega).$$

The space $L_2(\Omega)$ can be interpreted as $H^0(\Omega)$. Therefore, Sobolev space $H^m(\Omega)$ is determined for any integer degree[27]. The generalized differentiation is the linear continuous operator from $H^m(\Omega)$ to $H^{m-1}(\Omega)$ (see Figure 8.6).

The most important Sobolev space is $H^1(\Omega)$. The scalar product and the norm are determined here by the equalities

$$(u, v) = \int_\Omega \left[u(x) v(x) + \sum_{i=1}^m \frac{\partial u(x)}{\partial x_i} \frac{\partial v(x)}{\partial x_i}\right] dx,$$

$$\|u\| = \sqrt{\int_\Omega \left[|u(x)|^2 + \sum_{i=1}^m \left|\frac{\partial u(x)}{\partial x_i}\right|^2\right] dx}.$$

We consider often the space $H_0^1(\Omega)$ with the scalar product

$$(u, v) = \int_\Omega \sum_{i=1}^m \frac{\partial u(x)}{\partial x_i} \frac{\partial v(x)}{\partial x_i} dx,$$

$$\|u\| = \sqrt{\int_\Omega \sum_{i=1}^m \left|\frac{\partial u(x)}{\partial x_i}\right|^2 dx}.$$

The space $H^{-1}(\Omega)$ is adjoint to $H_0^1(\Omega)$, besides

$$\langle u, v \rangle = \int_\Omega u(x)v(x)dx \quad \forall u \in H^{-1}(\Omega), \ v \in H_0^1(\Omega).$$

Particularly, for the one-dimensional case we have the set $\Omega = (a, b)$. The scalar product of the space $H^1(a, b)$ is determined by the equality

$$(u, v) = \int_\Omega \left[u(x)v(x) + \frac{du(x)}{dx} \frac{dv(x)}{dx}\right] dx.$$

The norm of this space is

$$\|u\| = \sqrt{\int_\Omega |u(x)| + \left|\frac{\partial u(x)}{\partial x}\right|^2 dx}.$$

The space $H_0^1(a, b)$ has the scalar product

$$(u, v) = \int_\Omega \frac{du(x)}{dx} \frac{dv(x)}{dx} dx$$

and the norm

$$\|u\| = \sqrt{\int_\Omega \left|\frac{\partial u(x)}{\partial x}\right|^2 dx}.$$

We consider now the convergence of sequences of Sobolev spaces. Determine embedding properties.

Definition 8.10 *Embedding of a Banach space X to a Banach space Y is* **continuous,** *if from $u_k \to u$ in X it follows $u_k \to u$ in Y. This embedding is* **compact** *if $u_k \to u$ strongly in Y whenever $u_k \to u$ weakly in X.*

At first, embedding of the space $W_p^m(\Omega)$ to $W_q^s(\Omega)$ (particularly, of $H^m(\Omega)$ to $H^s(\Omega)$) is continuous for all $p > q \geq 1$; that is from the convergence $u_k \to u$ in $W_p^m(\Omega)$ it follows $u_k \to u$ in $W_q^s(\Omega)$. Note that embedding of the space $L_2(\Omega)$ to $H^{-1}(\Omega)$ is continuous too.

Theorem 8.2 (***Sobolev embedding theorem***[28]) *Let Ω be a bounded set of $\mathbb{R}^n$ with regular enough boundary. Then embedding $W_p^m(\Omega)$ in $W_q^s(\Omega)$ is continuous for $0 \le s \le m$ and $1/p-(s-k))/n \le 1/q < 1$. If $1/p-(s-k))/n < 0$, then embedding $W_p^m(\Omega)$ in $C^s(\overline{\Omega})$ is continuous.*

Particularly, the space $H^1(\Omega)$ (and $H_0^1(\Omega)$ too) has continuous embedding to $L_q(\Omega)$, if $q \le 2n/(n-2)$ for $n > 3$, to $L_q(\Omega)$ with arbitrary value q for $n = 2$, and to $C(\overline{\Omega})$ for $n = 1$.

Remark 8.32 The functions from the space $H^1(a,b)$ are continuous. Then the product of the functions of this space belongs to $H^1(a,b)$. Therefore, the multiplication is the operation here.

Theorem 8.3 (***Rellich–Kondrashov theorem***[29]) *Under the regular enough boundary of the set Ω, embedding of the space $W_p^m(\Omega)$ to $W_p^{m-1}(\Omega)$ is compact if $p \ge 1$, ≥ 1 The regularity supposition of the boundary is not necessary for compact embedding of the space $\dot{W}_p^m(\Omega)$ to $\dot{W}_p^{m-1}(\Omega)$.*

These results will be used for the analysis of the mathematical models of the heat transfer phenomenon.

8.5   Conclusions

1. The determination of mathematical models of physical phenomena requires the passage to the limit in the balance relations.

2. The justification of the passage to the limit is based on the Cauchy criterion, according to which any fundamental sequence converges.

3. The Cauchy criterion is applicable for the complete spaces only.

4. Most spaces are incomplete.

5. A standard example of an incomplete space is the set of rational numbers.

6. The divergent fundamental sequence of rational numbers becomes convergent if it is considered on the set of real numbers.

7. The set of real numbers has several interpretations.

8. By Cantor's interpretation, real numbers are understood as classes of equivalent fundamental sequences of rational numbers.

9. Cantor's interpretation is constructive, because of the possibility to approximate any real number with rational numbers with any degree of exactness.

10. The technique of Cantor extends the incomplete metric space to the

set of all classes of equivalent fundamental sequences of the original space that are the sequential objects.

11. Each element of the completion can by approximated by elements of the initial space with an arbitrary degree of exactness.

12. The analysis of the classes of equivalent fundamental sequences and the transfer from them of different properties of the original space to its completion are the basis of the sequential method.

13. The typical examples of the sequential objects are the real and p-adic numbers, the sequential controls, and the distributions.

14. Like real numbers, the theory of distributions has several interpretations, including the sequential one.

15. By sequential interpretation, each distribution can by approximated by smooth functions with an arbitrary degree of exactness.

16. The distribution theory is the basis of the Sobolev spaces which are the most functional spaces of mathematical physics.

We noted the effectiveness of the sequential approach in analyzing a wide class of difference mathematical problems. Now we have to try to apply it to solve the initial problem of justifying the procedure for constructing a mathematical model. Moreover, taking into account that the theory of distributions, which has a sequential interpretation, is used in an essential way in the analysis of the generalized model of mathematical physics problems, it is hoped that in this way we will be able to substantiate a generalized approach in mathematical physics. Using the well-known relations between the classical and generalized approaches, one can try to justify the classical approach too.

Notes

[1]The set is **closed**, if this is a complement to an open set. The open set is the general topological property that is determined by axioms. Particularly, the topology of a set is the set of all open sets here; see, for example, [6], [7], [26], [83], [173].

[2]The subset of the set of real numbers (and of the Euclid space which is **compact**, if this is closed and bounded. The compactness is one of the general properties of the set of topological spaces theory; see for example, [6], [7], [26], [83], [173].

[3]The test functions were determined by *Laurent Schwartz* in in the 1940s.

[4]The proof of the completeness of the space $D(\Omega)$ is given, for example, in [147].

[5]The example 8.3 of non-complete metrisable space is considered in [147].

[6]The distributions as linear continuous functionals on the space of infinite differentiable functions are definite by *Laurent Schwartz*. The complete enough distribution theory is described, for example, in [11], [29], [43], [95], [147], [158], [196].

[7]The relation between the local integrable functions and the corresponding distribution is described in detail in [71]. The analysis of linear partial differential operators I, Grundl. Math. Wissenschaft., 256, Springer, doi:10.1007/978-3-642-96750-4, ISBN 3-540-12104-8, MR 0717035.

[8]The δ–function and a class of distribution can be described on the basis of the Stieltjes measure; see, for example [142], [64].

[9]The Schwartz example about non-associativity of distribution multiplication was determined by *Laurent Schwartz* in 1954 [157]; see also [11], [47].

[10]The properties of the distribution $1/x$ are described in.

[11]The discussion of distribution multiplication is given, for example, in [47]. The first results in the field of distribution multiplications were obtained by *Y. Hirata* and *H. Ogata* [69] and *J. Mikusinski* [11] in the 1950s. The serious enough theory of distribution multiplication is described by *J.F. Colombeau* [38], [39]; see also [29], [75], [76], [92], [154], [186].

[12]Particularly, matrix multiplication and multiplication of the **quaternion** that is a generalization of the complex numbers are non-commutative [198].

[13]Sets with non-associative multiplication still occur in practice. Such, for example, are **octonions** belonging to the class of hypercomplex numbers. Each octonion is characterized by eight real numbers, just as a complex number is characterized by two, and the quaternion by four real numbers [200].

[14]The generalized derivative was determined by *Sergey Sobolev* in 1930s.

[15]The possibility of the approximation of generalized functions by smooth enough functions was determined by *Sergey Sobolev* in 1935. The proof of Theorem 8.1 that guarantees the approximation of any distribution by infinite differentiable functions is given, for example, in [11], [196].

[16]See [162].

[17]The topological space is **Hausdorf**, **separated** or T_2 **space** is a topological space in which distinct points have disjoint neighborhoods; see [26], [83]. The easiest example of the non-Hausdorf space is the **antidiscrete space**. Each point of this space has a unique neighborhood. This is whole space. Then any sequence of this space is convergent to all points of this space. Of course, the convergence does not apply in this situation.

[18]The theory of uniformly spaces is described in [26].

[19]The sequential definition of the distributions was given by *Jan Mikusinski* in 1955 [127]. The properties of this form of distributions are described in [11].

[20]The derivative of the sequential distributions was determined by *Jan Mikusinski* in 1955 [127].

[21]About the distribution multiplication problem see, for example, [11], [38], [39], [47], [92], [186].

[22]The theory of spaces of integrable functions is considered in the standard textbook on functional analysis; see, for example, [71], [81], [91], [95], [142], [147], [177], [203].

[23]About the Lebesgue measure, measurable functions, Lebesgue integral see, for example, [71], [91], [142].

[24]Each countable set has zero measure. However, there exist the infinite non-countable sets with zero measure. This is true, for example for the Cantor set, see [57].

[25]For example the integral of the Dirichlet function that is equal to one at the rational and equal to zero at the irrational is equal to zero. This measurable function is equivalent to the function that is equal to zero everywhere.

[26]The theory of **Sobolev spaces** is considered in [1], [21], [115], [124], [177].

[27]In reality, it is possible to determine the Sobolev spaces of non-integer degree too; see [115].

[28]The Sobolev embedding theorem was proved by *Sergey Sobolev* [177]; see also [133].

[29]The Rellich–Kondrashov theorem is considered in [1], [115], [124].

Part IV

Sequential models

The final part of the book contains a unique chapter. A constructive definition of the sequential model is given here by means of the sequential method described earlier. This result is used also for the substantiation of the generalized and classic models of stationary heat transfer.

Chapter 9

Sequential models of mathematical physics phenomena

This is our final chapter (see Figure 9.1). We would like to determine a mathematical model of the heat transfer phenomenon under corresponding suppositions. Using a known physical law, we obtain the balance relations (see Chapter 1)

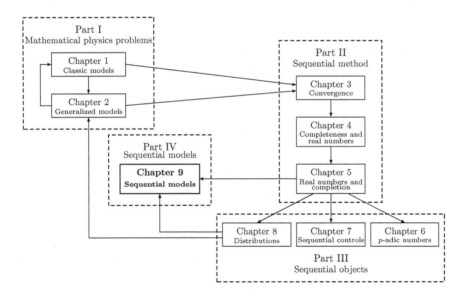

FIGURE 9.1: Structure of the book.

$$\frac{q(x) - q(x+h)}{h} = \frac{1}{h} \int\limits_{x}^{x+h} f(\xi)d\xi, \tag{9.1}$$

$$q(x) = -k(x)\frac{u(x) - u(x-h)}{h} \tag{9.2}$$

on the selected elementary interval $(x, x+h)$ of the given set $(0, L)$. If we pass to the limit here as $h \to 0$, then we could obtain the classic mathematical model that consist of the boundary problem (see Chapter 1)

$$\frac{d}{dx}\left[k(x)\frac{du(x)}{dx}\right] = f(x), \ x \in (0, L); \ \ u(0) = 0, \ u(L) = 0.$$

If we multiply the equality (9.1) by an arbitrary function λ from the Sobolev space $H_0^1(0, L)$ and integrate by x, then after passing to the limit we could determine the generalized mathematical model (see Chapter 2)

$$-\int\limits_{0}^{L} k(x)\frac{d\lambda(x)}{dx}\frac{du(x)}{dx}dx = \int\limits_{0}^{L} \lambda(x)f(x)dx \ \ \forall \lambda \in H_0^1(0, L).$$

For both cases, we have the same serious problem. This is the substantiation of the convergence. The desired result can be obtained if the state function u has a priori properties. It necessary that this function belongs to the set of twice continuously differentiable functions for the classic case and to the Sobolev space $H_0^1(0, L)$ for the generalized one. In principle, we could prove these results under some suppositions with respect to the known functions k and f by the differential equation theory after obtaining mathematical models. However, we have no properties of the state function before the definition of these mathematical models. Therefore, the determination of these mathematical models is not clear.

Then we consider the general definition of the convergence. The sequence tends to a limit if its elements with large enough numbers are close enough to this limit (see Chapter 3). Unfortunately, we cannot use this definition for proving the convergence, because for the practical situation we know, as a rule, the elements of the sequence only. However, we do not know the limit. We do not know even if the considered sequence is convergent or not.

The general practical means of the substantiation of passage to the limit is the Cauchy criterion (see Chapter 3). Particularly, if a sequence is fundamental, then it has a limit. This result is applicable, in principle, for the practical situation, because the definition of the fundamentality of sequences uses the elements of the sequence only. Unfortunately, this result is applicable for the complete spaces only (see Chapter 4). Unfortunately, a majority of spaces are non-complete. Thus, we have the serious problem of proving the convergence for the general case.

The reliable method of analysis of the convergence for non-complete spaces is based on the technique of Cantor's definition of real numbers (see Chapter 4). Particularly, the space of rational numbers is non-complete. However, each non-convergent fundamental sequence of rational numbers determines an irrational real number. Besides, different fundamental sequences can determine the same real number. These sequences are called equivalent. Therefore, each real number by Cantor is the equivalence class of fundamental sequences of rational numbers. It is very important that each real number can be approximate by rational numbers. Besides, the space of real numbers is complete.

This magnificent result can be extended to the general case. Particularly, a fundamental sequence of the arbitrary metric space can be non-convergent. However, it has a limit on the extension of the given space, which is called its completion (see Chapter 5). The initial metric space is isometric to a subspace of this completion that is dense there. Therefore, each element of the completion can be approximated by elements of the initial spaces. Besides, the completion is the complete metric space.

Thus, we have the forceful method of the substantiation of passing to the limit. At first, we try to prove that our sequence is fundamental using the elements of this sequence only. Then this sequence has a limit. Maybe this limit is an element of the given space; and this is the easy case. Maybe this limit is an element of its completion only. This case is more difficult. However, we know the determination of the completion and the possibility of the approximation of this limit by elements of the initial space. This is the general idea of the sequential method.

At first, we applied the sequential method for the definition of p-adic numbers (see Chapter 6). This is a special numerical class. The set of p-adic numbers is the completion of the set of rational numbers with respect to the p-adic metric.

Another example of the sequential object is the sequential control (see Chapter 7). The optimization control problems are often insolvable. However, the exact lower bound of the minimizing functional exists whenever this functional is lower bounded on the set of admissible controls. We determine the sequential controls as equivalence classes of the sequences of usual controls. Then we determine the sequential extension of the given optimization problem as a problem of minimization, a special functional on the set of sequential controls. This problem is solvable without fail. Besides, its solution is determined by minimizing sequences of the initial optimization problem. Therefore, we can choose an element of such a sequence with a large enough number as an approximate solution of the initial problem.

Finally, we apply the sequential technique for the determination of distributions. This application is very important, because the distribution theory is the basis of the generalized models of mathematical physics problems. One can determine the set of distributions as the completion of the space of infinite differential functions. Therefore, each distribution can be approximated by smooth functions.

Now we try to use the sequential method for the correct determination of a new form of mathematical model of the considered phenomenon. We determine the sequential model of the heat transfer phenomenon. Then we prove the existence of the generalized and classic state of the considered phenomenon.

9.1   Sequential model of the heat transfer phenomenon

We return to the consideration of the stationary one-dimensional heat transfer phenomenon in a non-homogeneous body under a source of heat with zero temperature at the ends of the body. Divide the given set $\Omega = [0, L]$ by M equal parts with the step $h = L/M$. Determine the points $x_i = ih$, where $i = 0, ..., M$. Consider the elementary intervals

$$\Omega_i = (x_{i-1}, x_i), \; i = 1, ..., M.$$

Determine the set

$$\Gamma_h = \{\Omega_1, ..., \Omega_M\}.$$

The **grid function** on the set Γ_h is the vector of the order $M+1$ with indexes $0, 1, ..., M$.

The state of the system in the set Ω_i satisfies the balance relations (9.1), (9.2) with $x = x_i$

$$\frac{q(x_i) - q(x_i + h)}{h} = \frac{1}{h} \int\limits_{x_i}^{x_i+h} f(\xi)d\xi, \tag{9.3}$$

$$q(x_i) = -k(x_i)\frac{u(x_i) - u(x_i - h)}{h}. \tag{9.4}$$

Add also the boundary conditions

$$u(0) = 0, \; u(L) = 0. \tag{9.5}$$

We used, in principle, the analogical equalities for obtaining the classic and generalized mathematical models of the considered phenomenon (see Chapter 1 and Chapter 2).

Determine the standard difference operators

$$\delta_{\bar{x}} : \mathbb{R}^{M+1} \to \mathbb{R}^M, \; \delta_x : \mathbb{R}^{M+1} \to \mathbb{R}^M$$

by the formulas (see Chapter 1)

$$\delta_{\bar{x}} u_i = \frac{u_i - u_{i-1}}{h}, \; i = 1, ..., M;$$

$$\delta_x u_i = \frac{u_{i+1} - u_i}{h}, \ i = 0, ..., M - 1.$$

Then we transform the balance relations (9.3), (9.4) to the **difference equations**

$$\delta_x (k_i \delta_{\bar{x}} u_i) = f_i, \ i = 1, ..., M - 1, \tag{9.6}$$

where

$$u_i = u(x_i), \ k_i = k(x_i), \ f_i = \frac{1}{h} \int\limits_{x_i}^{x_i+h} f(\xi)d\xi.$$

From the boundary conditions (9.5), it follows that

$$u_0 = 0, \ u_M = 0. \tag{9.7}$$

The system of linear algebraic equations (9.6), (9.7) with triangle matrix is the **discrete model** of the system, and its solution is the **discrete state** of the system. We obtain it from the known physical law directly without using any mathematical hypothesis. Then we can solve[1] the system (9.6), (9.7) and find all values u_i, which determine the grid function

$$\bar{u}_h = \{u_0, u_1, ..., u_M\}.$$

Now we determine the linear interpolation of the grid function (see Figure 9.2)

$$u_h(x) = u_{i-1} + (x - x_{i-1})\delta_{\bar{x}} u_i, \ x \in \Omega_i, \ i = 1, ..., M. \tag{9.8}$$

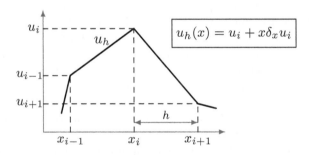

FIGURE 9.2: Linear interpolation of the grid function.

Consider a linear topological space H of functions with the domain Ω. The sequence of linear interpolations $\{u_h\}$ of H is **fundamental** if $(u_h - u_{h'}) \to 0$ in H as $h \to 0$. Determine the **equivalence** of the fundamental sequences on the set H such that the sequence $\{u_h\}$ and $\{v_h\}$ are equivalent if $(u_h - v_h) \to 0$ in H as $h \to 0$. Now we can give the general definition[2].

TABLE 9.1:	Sequential states and models

object of analysis	sequential state	sequential model	approximation
heat transfer	sequential state $[u_h]$	sequential model $\{u_h\}$	interpolation u_h
number theory	irrational number	sequence of rational numbers	rational number
number theory	p-adic number	sequence of rational numbers	rational number
metric spaces	point of the completion	sequence of elements of initial metric space	element of initial metric space
optimization control problem	sequential control	sequence of admissible controls	admissible control
distribution theory	singular distribution	sequence of smooth functions	smooth function

Definition 9.1 *If the sequence $\{u_h\}$ of H is fundamental, then this is the **sequential model** of the system, and the corresponding equivalence class $[u_h]$ is called the **sequential state**.*

Remark 9.1 More exactly, we have *H-sequential model* and *H-sequential state* of the system, because these notions depend on the space H of sequence fundamentality.

The determination of the sequential model and the sequential state is correct. Indeed, for any step h we can determine the solution of the difference system (9.6), (9.7) which is a grid function. Then we determine its linear interpolation u_h by the formula (9.8). Therefore, we have the sequence $\{u_h\}$. If we prove its fundamentality with respect to a space H, then we obtain the sequential state of the system without any a priori mathematical hypothesis with respect to the state of the system. This is an analogue of the Cantor real number, the p-adic number, the sequential distribution, the element of the completion for the metric spaces, and the sequential control (see Table 9.1). Then we choose the function u_h with small enough step h as an approximate solution of the considered problem. This is the analogue of the approximation of the irrational number by a rational number, the distribution by a smooth function, the element of completion by an element of the initial metric space, and the sequential control by an admissible one.

Note that we work, in reality, with rational numbers and smooth functions that are approximations of irrational numbers and distributions, and not with irrational numbers and distributions themselves. Analogically, we have the sequential model $[u_h]$, but we apply its approximation u_h with small enough step h.

Thus, we determine three different forms of mathematical models by the same balance relations. However, we have the same difference equations (9.6), (9.7) for finding the approximate solution of the system for all cases. Therefore, it could give us the approximation of classic, generalized, and sequential models.

9.2 Justification of sequential modeling

We have so far determined the sequential model. However, we did not substantiate it. At first, it is necessary to formulate some additional results of the discrete functions theory. This is the basis of the justification of the *finite difference method*[3]. Consider the properties of the discrete spaces[4]. Denote by V_h the set of all grid functions. This, in reality, is the very known set $\mathbb{R}^{M+1}$. However, we determine there the scalar product

$$\left(\overline{u}_h, \overline{v}_h\right)_h = h \sum_{i=1}^{M} u_i v_i.$$

The corresponding norm is determined by the equality

$$\left|\overline{v}_h\right|_h = \sqrt{\left(\overline{v}_h, \overline{v}_h\right)_h}.$$

Determine here also another norm

$$\left\|\overline{v}_h\right\|_h = \sqrt{h \sum_{i=1}^{M} \left(\delta_{\overline{x}} v_i\right)^2}$$

and the norm of the adjoint space

$$\left\|\overline{v}_h\right\|_{*h} = \sup_{\left\|\overline{u}_h\right\|_h = 1} \left|\left(\overline{u}_h, \overline{v}_h\right)_h\right|.$$

Remark 9.2 For any linear normalized space V the norm of adjoint space V', i.e. the set of all linear continuous functionals, is determined by the equality

$$\|v\|_{V'} = \sup_{\|u\|_V = 1} |\langle u, v \rangle|,$$

where $\langle u, v \rangle$ is the value of the linear continuous functional u at the point v. This is true, for example, for the space $H^{-1}(0, L)$ that is adjoint to $H_0^1(0, L)$ (see Chapter 8). By the *Riesz theorem*, for any Hilbert space each linear continuous functional can be determined by the scalar product[5].

Note the inequality

$$\left|\left(\overline{u}_h, \overline{v}_h\right)_h\right| \leq \left\|\overline{u}_h\right\|_h \left\|\overline{v}_h\right\|_{*h} \; \forall \overline{u}_h, \overline{v}_h \in V_h. \tag{9.9}$$

The set V_h with norms $|\cdot|_h$, $\|\cdot\|_h$, and $\|\cdot\|_{*h}$ is the discrete analogue of the spaces $L_2(0, L)$, $H_0^1(0, L)$, and $H^{-1}(0, L)$.

Remark 9.3 The formula (9.9) is the partial case of the inequality

$$|\langle u, v \rangle| \leq \|u\|_V \|v\|_{V'} \; \forall u \in V, v \in V'$$

for all linear normalized spaces V.

For any grid function $\bar{v}_h$ determine its linear interpolation (see Figure 9.2)

$$v_h(x) = v_{i-1} + (x - x_{i-1})\delta_{\bar{x}}v_i, \ x \in \Omega_i, \ i = 1, ..., M$$

and its piecewise constant interpolation (see Figure 9.3)

$$v^h(x) = v_i, \ x \in \Omega_i, \ i = 1, ..., M.$$

These functions are determined on the set Ω and depend from the step h. The operator that maps the grid function $\bar{v}_h$ to the corresponding function with the domain Ω, for example, its interpolations v_h and v_h, is called the **extension operator**.

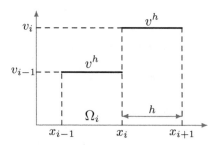

FIGURE 9.3: Piecewise constant interpolation of the grid function.

Now return to the consideration of the discrete model (9.6), (9.7). Multiply the equality (9.6) by u_i and sum by i. We obtain

$$\sum_{i=1}^{M-1} \left[u_i \delta_x \left(k_i \delta_{\bar{x}} u_i \right) \right] = \sum_{i=1}^{M-1} u_i f_i.$$

Using the formula of **summing by parts** (see the approximation of the generalized model, Chapter 2), for all grid function such that v_M is zero we have

$$\sum_{i=1}^{M-1} \left(\delta_x v_i \right) \lambda_i = \frac{1}{h} \sum_{i=1}^{M-1} \left(v_{i+1} - v_i \right) \lambda_i = - \sum_{i=1}^{M} v_i \left(\delta_{\bar{x}} \lambda_i \right).$$

Then we transform the previous equality to

$$-\sum_{i=1}^{M} k_i \left(\delta_{\bar{x}} u_i \right)^2 = \sum_{i=1}^{M} u_i f_i. \tag{9.10}$$

Further transformations require some restrictions on the given functions k and f. Obviously, the function k that is the coefficient of thermal conductivity is positive in the physical sense. Suppose the following inequality holds

$$k(x) \geq k_0, \ x \in (0, L), \tag{9.11}$$

where k_0 is a positive constant. From the equality (9.10) and the definition of the discrete norms, it follows that

$$k_0 \|\bar{u}_h\|_h^2 \leq \left| (\bar{u}_h, \bar{f}_h)_h \right|.$$

Using the inequality (9.9), we have

$$k_0 \|\bar{u}_h\|_h \leq \|\bar{f}_h\|_{*h}. \tag{9.12}$$

Differentiate the equality (9.8). We get

$$\frac{du_h(x)}{dx} = \delta_{\bar{x}} u_i, \ x \in \Omega_i, \ i = 1, ..., M.$$

Then we have

$$\|u_h\|_{H_0^1}^2 = \int_0^L \left[\frac{du_h(x)}{dx} \right]^2 dx = \sum_{i=1}^M \int_{x_i}^{x_{i+1}} \left(\delta_{\bar{x}} u_i \right)^2 dx = h \sum_{i=1}^M \left(\delta_{\bar{x}} u_i \right)^2 = \|\bar{u}_h\|_h^2.$$

Suppose the function f satisfies the condition $f \in H^{-1}(0, L)$. Then the following inequality holds

$$\|\bar{f}_h\|_{*h} \leq c,$$

where a positive constant c does not depend on h. From the formulas (9.12), the boundedness of the sequence $\{u_h\}$ in $H_0^1(0, L)$ follows. Using the Banach-Alaoglu theorem (see Chapter 8), extract a subsequence that is weakly convergent in $H_0^1(0, L)$ as $h \to 0$. This is the weak convergence of the derivatives of the functions u_h in the space $L_2(0, L)$. The convergence is stronger property than the fundamentality. Therefore, the sequence $\{u_h\}$ (more exactly, its subsequence) is fundamental with respect to the weak topology of the Sobolev space. Thus, the class $[u_h]$ is the sequential state of the considered system, and the corresponding equivalence class is its sequential model. We proved the following result[6]

Theorem 9.1 *If the function k is lower bounded by a positive constant, and the function f belongs to the space $H^{-1}(0, L)$, then the sequence $\{u_h\}$ that is the linear interpolation of the grid function $\bar{u}_h$, determined by the equalities (9.6), (9.7), is the sequential model of the stationary heat transfer phenomenon with respect to the space $H_0^1(0, L)$ with weak topology. The corresponding equivalence class is the sequential state of the system.*

Remark 9.4 We determine this result for a subsequence of $\{u_h\}$, not for the whole sequence. However, using the analogical technique, we can extract a subsequence from the arbitrary subsequence of $\{u_h\}$ that will be a sequential model of the system.

Remark 9.5 We cannot guarantee the uniqueness of the sequential state here. Therefore, the different subsequences of $\{u_h\}$ can tend to the different sequential states.

Thus, we have achieved the desired goal by constructing a mathematical model of the stationary heat transfer process without a priori constraints on the solution of the problem. The determined sequential state is the analogue of the Cantor real number, the p-adic number, the sequential control, and the element of the completion of a metric space (see Table 9.1).

It is important that, starting from the discrete balance relations after the passage to the limit we determined the continuous classical and generalized models. Besides, for the practical solution of the problem these continuous models are again discretized. In this case, all three stages of research are clearly identified and consistently implemented: the determination of the mathematical model, the proof of the existence of the corresponding state, and practical realization with the proof of the convergence of the algorithm for solving the problem based on the discrete model (see Figure 9.4). By the sequential approach, these steps are carried out simultaneously. Indeed, the discrete model is a direct consequence of physical laws, and the fundamentality of the sequence of interpolations of the grid function can serve both as a way to construct the sequential model, as the basis for the existence of the sequential state, and the basis for proving the algorithm for finding it (see next section).

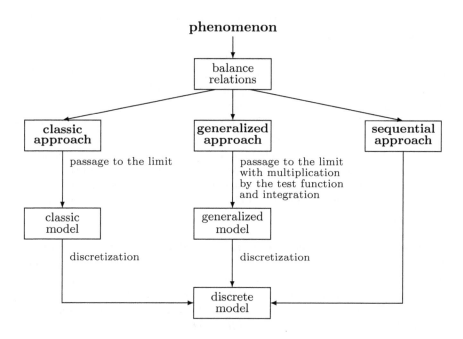

FIGURE 9.4: Analysis of the problem by the different approaches.

Note that in this case it is possible not only to restrict ourselves to the sequential model, but to derive a generalized model from it, thereby providing a justification for the generalized approach.

9.3   Generalized model of the heat transfer phenomenon

The existence of a certain connection between the sequential and generalized models is indicated by the fact that in both cases the state of the system is associated with the same Sobolev space $H_0^1(0, L)$. Note also the fact that earlier we established not only the fundamentality, but also the weak convergence of the sequence $\{u_h\}$ (more exactly, of its subsequence). Then there exists a function u of this space such that $u_h \to u$ weakly in $H_0^1(0, L)$. The question arises, what are the properties of this limit u? In general, we cannot answer it. However, under additional conditions one can pass to the limit in a discrete model and determine a relation with respect to the function u.

Consider a smooth enough function λ on the set Ω with zero values on its boundary. Denote by λ_i the value of λ at the point x_i. Multiply the i-th equality (9.6) by the number λ_i and sum by the index i. We get

$$\sum_{i=1}^{M-1} \left[\delta_x\left(k_i \delta_{\bar{x}} u_i\right) \lambda_i \right] = \sum_{i=1}^{M-1} f_i \lambda_i.$$

Using the formula of summing by parts, for any grid function with components g_i such that g_M is zero we have

$$\sum_{i=1}^{M-1} \left(\delta_x g_i\right) \lambda_i = \frac{1}{h} \sum_{i=1}^{M-1} \left(g_{i+1} - g_i\right) \lambda_i = \frac{1}{h} \sum_{i=1}^{M-1} g_i \left(\lambda_i - \lambda_{i-1}\right) = -\sum_{i=1}^{M-1} g_i \left(\delta_{\bar{x}} \lambda_i\right).$$

Then we can transform the previous equality to

$$-\sum_{i=1}^{M} \left(k_i \delta_{\bar{x}} u_i \delta_{\bar{x}} \lambda_i\right) = \sum_{i=1}^{M} f_i \lambda_i. \tag{9.13}$$

Find the derivative

$$\frac{d\lambda_h(x)}{dx} = \delta_{\bar{x}} \lambda_i, \ x \in \Omega_i, \ i = 1, ..., M,$$

where λ_h is the linear interpolation of the grid function $\bar{\lambda}_h$. Note that the piecewise linear functions u_h and λ_h are continuous and also differentiable on each open interval Ω_i. Then we find the integral

$$\int_0^L k^h \frac{du_h}{dx} \frac{d\lambda_h}{dx} dx = \sum_{i=1}^{M} \int_{\Omega_i} k^h \frac{du_h}{dx} \frac{d\lambda_h}{dx} dx = h \sum_{i=1}^{M} \left(k_i \delta_{\bar{x}} u_i \delta_{\bar{x}} \lambda_i\right),$$

where k^h is the piecewise-constant interpolation of the grid function $\bar{k}_h$ (see Figure 9.3).

Find the grid function $\bar{g}_h$ from the equalities

$$g_0 = 0, \quad \delta_x g_i = f_i, \quad i = 1, ..., M-1.$$

Using the formula of integration by part, we have

$$\int_0^L g^h \frac{d\lambda_h}{dx} dx = \sum_{i=1}^M \int_{\Omega_i} g^h \frac{d\lambda_h}{dx} dx = h \sum_{i=1}^M \left(g_i \delta_{\bar{x}} \lambda_i \right) =$$

$$-h \sum_{i=1}^M \left(\delta_x g_i \lambda_i \right) = -h \sum_{i=1}^M f_i \lambda_i,$$

where g^h is the point-constant interpolation of the grid function $\bar{g}_h$.

By the previous equalities, the formula (9.13) is transformed to

$$\int_0^L k^h \frac{du_h}{dx} \frac{d\lambda_h}{dx} dx = \int_0^L g^h \frac{d\lambda_h}{dx} dx. \tag{9.14}$$

Pass to the limit here as $h \to 0$.

Denote by f^h the generalized derivative of the function g^h. We have $f^h \to f$ in $H^{-1}(0, L)$ as $h \to 0$. This is equivalent to the convergence $g^h \to g$ in $L_2(0, L)$, where the generalized derivative of g is equal to f.

Determine the convergence

$$\frac{d\lambda_h}{dx} \to \frac{d\lambda_h}{dx} \quad \text{in} \quad L_2(0, L).$$

We have the inequality

$$\left| \int_0^L \left(g^h \frac{d\lambda_h}{dx} - g \frac{d\lambda_h}{dx} \right) dx \right| \leq \left| \int_0^L (g^h - g) \frac{d\lambda_h}{dx} \right| + \left| \int_0^L g \left(\frac{d\lambda_h}{dx} - \frac{d\lambda}{dx} \right) dx \right| \leq$$

$$\left\| g^h - g \right\|_{L_2} \left\| \frac{d\lambda_h}{dx} \right\|_{L_2} + \left\| g \right\|_{L_2} \left\| \frac{d\lambda_h}{dx} - \frac{d\lambda}{dx} \right\|_{L_2}.$$

By the convergence of the sequence $\{g^h\}$, we get

$$\int_0^L g^h \frac{d\lambda_h}{dx} dx \to \int_0^L g \frac{d\lambda}{dx} dx.$$

Now we obtain

$$\int_0^L g \frac{d\lambda}{dx} dx = g(L)\lambda(L) - g(0)\lambda(0) - \int_0^L \lambda \frac{dg}{dx} dx = - \int_0^L \lambda f dx.$$

Thus, we prove the convergence

$$\int_0^L g^h \frac{d\lambda_h}{dx} dx \to - \int_0^L \lambda f dx.$$

It is necessary to have the additional properties of the function k for passing to the limit at the left-hand side of the equality (9.14). Under its continuity we have the convergence $k_h \to k$ in $[0, L]$. Analyze the integral

$$\left| \int_0^L \left(k^h \frac{du_h}{dx} \frac{d\lambda_h}{dx} - k \frac{du}{dx} \frac{d\lambda}{dx} \right) dx \right| \leq \left| \int_0^L (k^h - k) \frac{du_h}{dx} \frac{d\lambda_h}{dx} dx \right| +$$

$$\left| \int_0^L k \left(\frac{d\lambda_h}{dx} - \frac{d\lambda}{dx} \right) \frac{du}{dx} dx \right| + \left| \int_0^L k \left(\frac{du_h}{dx} - \frac{du}{dx} \right) \frac{d\lambda}{dx} dx \right|.$$

By definition of the weak convergence of the sequence $\{u_h\}$ in $H_0^1(0, L)$, we have

$$\int_0^L \left(\frac{du_h}{dx} - \frac{du}{dx} \right) \mu dx \to 0$$

for all $\mu \in L_2(0, L)$.

The following inequality holds

$$\int_0^L \left| k \frac{d\lambda}{dx} \right|^2 dx \leq \max_{x \in [0,L]} |k(x)|^2 \int_0^L \left| \frac{d\lambda}{dx} \right|^2 dx < \infty.$$

Choose μ equal to the product of k and the derivative of λ. Then we have the convergence

$$\int_0^L \left(\frac{du_h}{dx} - \frac{du}{dx} \right) k \frac{d\lambda}{dx} dx \to 0.$$

Using the continuity of the function k, we get

$$\left| \int_0^L \left(k \frac{d\lambda_h}{dx} - \frac{d\lambda}{dx} \right) \frac{du_h}{dx} dx \right| \leq \max_{x \in [0,L]} |k(x)| \int_0^L \left| \left(\frac{d\lambda_h}{dx} - \frac{d\lambda}{dx} \right) \frac{du_h}{dx} \right| dx \leq$$

$$\|k\|_{C[0,L]} \left\| \frac{du_h}{dx} \right\|_{L_2} \left\| \frac{d\lambda_h}{dx} - \frac{d\lambda}{dx} \right\|_{L_2}.$$

By the boundedness of the sequence $\{u_h\}$ in $H_0^1(0, L)$ the corresponding sequence of derivatives is bounded in $L_2(0, L)$. Therefore, we have the convergence

$$\int_0^L k \left(\frac{d\lambda_h}{dx} - \frac{d\lambda}{dx} \right) \frac{du_h}{dx} dx \to 0.$$

Finally, from the inequality

$$\left| \int_0^L \left(k^h - k \right) \frac{d\lambda_h}{dx} \frac{du_h}{dx} dx \right| \leq \max_{x \in [0,L]} \left| k^h(x) - k(x) \right| \left| \int_0^L \frac{d\lambda_h}{dx} \frac{du_h}{dx} dx \right| \leq$$

$$\left\| k^h - k \right\|_{C[0,L]} \left\| \frac{d\lambda_h}{dx} \right\|_{L_2} \left\| \frac{du_h}{dx} \right\|_{L_2}$$

the convergence follows

$$\int_0^L k^h \frac{d\lambda_h}{dx} \frac{du_h}{dx} dx \to \int_0^L k \frac{d\lambda}{dx} \frac{du}{dx} dx.$$

Now we obtain

$$-\int_0^L k \frac{d\lambda}{dx} \frac{du}{dx} dx = \int_0^L \lambda f dx \quad \forall \lambda \in H_0^1(0, L). \tag{9.15}$$

There is, in reality, the generalized model of the considered phenomenon (see Chapter 2). Thus, the weak $H_0^1(0, L)$ limit u of the sequence $\{u_h\}$ as $h \to 0$ is the **generalized state** of the system. Therefore, we proved the following result[7].

Theorem 9.2 *Under the condition of Theorem 9.1 and the continuity of the function k, there exists the generalized state of the system that is the $H_0^1(0, L)$ weak limit of the linear interpolation of the discrete state.*

Thus, we not only give a correct determination of the mathematical model based on the sequential approach, but also justify the generalized approach. The existence of a limit of the sequence $\{u_h\}$, which is a generalized state of the system, is analogous to the convergence of the minimizing sequence to the optimal control and the fundamental sequences of rational numbers and infinite differentiable functions to their limits (see Table 9.2). These objects (generalized state, optimal control, rational number, smooth function) can be identified up to isomorphism with regular sequential objects (regular sequential state, sequentially optimal control, rational real number, regular generalized function). For this, it suffices to consider stationary sequences with a unique element equal to this limit. It follows that any generalized state of the system is in a certain sense its sequential state. However, for the proof of the solvability of the problem in the generalized sense, we needed an additional property of continuity of the function k that is not required for the construction of a sequential model. Thus, not every sequential state of the system coincides with its generalized state. These results are an analogue of the relation between the generalized and the classical states: the higher the degree of regularity of a state, the narrower the class of problems where it is realized (see Figure 9.5 and Figure 9.6).

TABLE 9.2: Sequential objects

application	class $[u_h]$	sequence $\{u_h\}$	value u_h	transition $u_h \Rightarrow u$	value u
physical phenomenon	sequential state	sequential model	approximate state	there exists the generalized state	generalized state
number theory	real number	approximating sequence of rational numbers	rational approximation of real number	real number is rational	rational number
number theory	p-adic number	approximating sequence of rational numbers	rational approximation of p-adic number	p-adic number is rational	rational number
distribution theory	distribution	approximating sequence of smooth functions	smooth approximation of distribution	distribution is regular	smooth function
metric space theory	element of the completion	fundamental sequence of initial elements	approximating object of the initial space	metric space is complete	element of the initial space
optimization control theory	sequentially optimal control	minimizing control sequence	approximate solution of the problem	optimization control problem is solvable	optimal control

Remark 9.6 We denote by S_c, S_g and S_s the set of pairs (k, f) that guarantee the existence of, respectively, the classical, generalized and sequential state of the system. The sets S_g and S_s are determined in Theorem 9.1 and Theorem 9.2. The set S_s will be described in the next section (see Theorem 9.4).

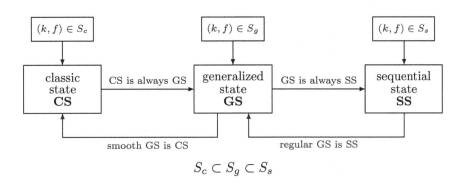

$$S_c \subset S_g \subset S_s$$

FIGURE 9.5: The higher the regularity of a state, the narrower the class of tasks.

Prove the uniqueness of the generalized state of the considered system.

Theorem 9.3 *Under the condition of Theorem 9.2, the generalized state of the considered system is unique.*

Proof. Suppose there exist two generalized states of the system u_1 and

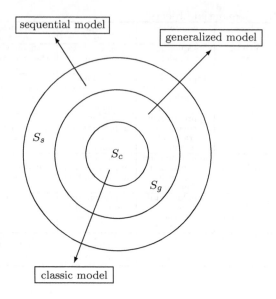

FIGURE 9.6: The degree of regularity of the parameters determines the type of the model.

u_2. Then we have

$$-\int_0^L k\frac{d\lambda}{dx}\frac{du_i}{dx}dx = \int_0^L \lambda f dx \ \ \forall \lambda \in H_0^1(0,L), \ i = 1,2.$$

The function $u = u_1 - u_2$ satisfies the equality

$$\int_0^L k\frac{d\lambda}{dx}\frac{du}{dx}dx = 0 \ \ \forall \lambda \in H_0^1(0,L).$$

Choose $\lambda = u$. We get

$$\int_0^L k\left(\frac{du}{dx}\right)^2 dx = 0.$$

From the inequality (9.11) and the definition of the norm of the space $H_0^1(0,L)$, it follows that

$$k_0\|u\|_{H_0^1} \leq 0.$$

Then $u = 0$, and the following equality holds $u_1 = u_2$. □

Remark 9.7 Since not every sequential state is generalized (see Figure 9.5), from the uniqueness of the generalized state, it does not follow that the sequential state is unique.

9.4 Classic model of the heat transfer phenomenon

Now we give the substantiation of the classic approach on the basis of the known relation between the classic and the generalized model. Prove the existence of the **classic state** of the considered system.

Theorem 9.4 *Suppose the function k is lower bounded by a positive constant k_0 and continuously differentiable in the interval $[0, L]$, and the function f is continuous there. Then there exists a unique twice differentiable function u such that*

$$\frac{d}{dx}\left[k(x)\frac{d}{dx}\right] = f(x), \ x \in (0, L), \tag{9.16}$$

$$u(0) = 0, \ u(L) = 0. \tag{9.17}$$

Proof. Under the given suppositions, from Theorem 9.3 the existence of a unique generalized state follows, i.e. the function of the space $H_0^1(0, L)$, which satisfies the equality (9.15). Using the continuity of the square integrable function, from the equality (9.16) the inclusion follows

$$\frac{d}{dx}\left[k(x)\frac{d}{dx}\right] \in L_2(0, L). \tag{9.18}$$

The function u as the generalized state of the boundary problem belongs to the Sobolev space $H^1(0, L)$. Therefore, the derivative u' of the function u belongs to $L_2(0, L)$. Using the continuity of the function k, determine the inclusion $ku' \in L_2(0, L)$. Now from the condition (9.18) it follows that $ku' \in H^1(0, L)$. By the Sobolev embedding theorem (see Chapter 8), we have the continuous embedding of the space $H^1(0, L)$ of one dimensional functions to the space of continuous function on the given interval. Then the product ku' is continuous. Consider the equality

$$u'(x) = \frac{k(x)u'(x)}{k(x)}.$$

Here in the denominator there is a function taking exclusively positive values, and, according to the conditions of the theorem, the following inequality holds $1/k(x) \le 1/k_0$. Therefore, we have the estimate

$$\|u'\|_{C[0,L]} = \max_{x \in (0,L)} |u'(x)| \le \frac{1}{k_0} \max_{x \in (0,L)} |k(x)u'(x)| = \frac{1}{k_0}\|ku'\|_{C[0,L]}.$$

Thus, the function u' is continuous, and the function u is continuously differentiable.

Using the continuity of the function f, from the equality (9.16) the continuity of the function $(ku')'$ follows. By differentiability of the functions k and u we have the continuity of the product $k'u'$. Then the function

$$ku'' = (ku')' - k'u'$$

TABLE 9.3:   System parameters and forms of states

result	set of parameter	k	f	form of state
Theorem 9.1	S_s	$k(x) \geq k_0 > 0$	$f \in H^1(0, L)$	sequential
Theorem 9.2	S_g	$k \in C[0, L]$ $k(x) \geq k_0 > 0$	$f \in H^1(0, L)$	generalized
Theorem 9.3	S_c	$k \in C^1[0, L]$ $k(x) \geq k_0 > 0$	$f \in C[0, L]$	classic

is continuous too. Consider the second derivative

$$u''(x) = \frac{k(x)u''(x)}{k(x)}.$$

The value at the denominator here is positive. Then the second derivative u'' is continuous; besides the following inequality holds

$$\|u''\|_{C[0,L]} = \max_{x \in (0,L)} |u''(x)| \leq \frac{1}{k_0} \max_{x \in (0,L)} |k(x)u''(x)| = \frac{1}{k_0}\|ku''\|_{C[0,L]}.$$

Using the obtained results, we conclude that the function u, satisfying the equality (9.15), belongs to the set $C^2[0, L]$. By Theorem 2.2, the twice continuously differentiable generalized state of the system is its classic state, i.e. the classic solution of the boundary problem (9.16), (9.17). Its uniqueness follows from the uniqueness of the generalized state that is the corollary of Theorem 9.3. □

Thus, by the sequential state of the system, under additional properties of its parameters, we can determine the generalized state of this system. In turn, imposing additional restrictions on the parameters of the problem, we can derive from here its classical state. The relationships between different types of system states (or solutions of the mathematical physics problem) are given in Table 9.3.

Remark 9.8 Suppose the function k satisfies the condition of Theorem 9.4, and the function f is square integrable. By Theorem 9.2 the considered problem has a unique solution from the space $H_0^1(0, L)$. The value $(ku')'$ belongs to the space $L_2(0, L)$, because this is equal to f by the equality (9.16). Then the sum $k'u' + ku''$ belongs to $L_2(0, L)$. From the continuity of the derivative k' and the inclusion $u' \in L_2(0, L)$ it follows that $k'u' \in L_2(0, L)$. Therefore, we have the condition $ku'' \in L_2(0, L)$, where the function k is continuous and has only positive value. Then we obtain $u'' \in L_2(0, L)$. Thus, under the considered supposition, our boundary problem has a solution from the space $H^2(0, L)$.

Remark 9.9 Boundary problems for the second order differential equations can have different properties[8]. The theory of the boundary problems for nonlinear differential equations is even more difficult[9].

Extremely important here is the fact that all three types of model are repelled from the same balance relations and output to the same discrete model. In this connection, the question arises: what type of solution is obtained when solving the difference equations (9.6), (9.7) in practice (see Figure 9.7)? Apparently, if it is not possible to establish even fundamentality for the sequence of interpolations uh of the grid function determined from difference equations, then we cannot guarantee finding even a sequential state of the system. If, however, the considered sequence is fundamental, but does not converge in the corresponding sense, then it is possible to obtain a sequential state of the system, but not to guarantee that it will become a generalized state. If under the conditions of the convergence of the sequence $\{u_h\}$ the corresponding limit is not sufficiently smooth, then the generalized state of the system that is not classical is realized. Finally, in the case of sufficient smoothness of the limit, it will already be the classical state of the system.

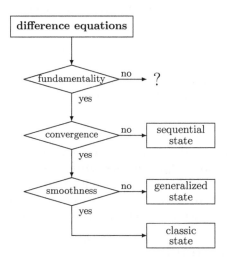

FIGURE 9.7: Practical realization of the difference equations.

Now we consider a formalization of the obtained results.

9.5   Models of mathematical physics problems

Consider a physical phenomenon, for example, the stationary heat transfer phenomenon (see Table 9.4). We have the process in the set Ω, for example, on the interval $[0, L]$. The given set is divided by the system Γ of measurable subsets $\{\Omega_\alpha\}$. The example of this system is the set of elementary interval $\Omega_i = (x_{i-1}, x_i)$. Denote by h_Γ the maximal measure of the set Ω_i. It was the step h for our situation.

Remark 9.10 α is a multiindex for the multidimensional case. The set Ω can also include a time interval if we consider a non-stationary system.

The discrete state of the system on the set Γ is described by a grid function $\bar{u}_\Gamma$. This is $\bar{u}_h$ for our case. Denote by V_Γ the set of all grid functions in Γ that is V_h for the considered system.

Remark 9.11 The state of the system can be described by a vector function.

The discrete state at the concrete elementary cell Ω_α is described by the equality

$$A_\alpha \bar{u}_\Gamma = 0$$

in accordance with any physical laws, where A_α is an operator. Determine for the stationary heat transfer phenomenon

$$A_i \bar{u}_h = \delta_x\left(k_i \delta_{\bar{x}} u_i\right) - f_i, \ i = 1, ..., M-1; \ \ A_0 \bar{u}_h = 0, \ A_M \bar{u}_h = 0.$$

Then determine the discrete state operator A_Γ on the set V_Γ such that A_α is its restriction on the set Ω_α. The discrete model of the system is the equality

$$A_\Gamma u_\Gamma = 0. \tag{9.19}$$

These are the equalities (9.6), (9.7) for the considered system.

Then we have a **prolongation operator** π that transforms the discrete state $\bar{u}_\Gamma$ to the function $u_\Gamma = \pi \bar{u}_\Gamma$, which is determined on the set Ω. This is an **approximate state** of the system. For example, this is the linear interpolation of the grid function, which satisfies the equality (9.8).

Remark 9.12 In reality, this will be an approximate state after the corresponding justification, see Theorem 9.1 and Theorem 9.2.

Consider the set of approximate solutions $\{u_\Gamma\}$ with respect to all possible partitions Γ of the set Ω.

Remark 9.13 The strict definition of the notions of the space that can be non-metric needs to use filter theory or net theory[10]. However, we shall, for simplicity, still refer to the set of approximate states $\{u_\Gamma\}$ as a sequence.

The convergence and the fundamentality of the sequence $\{u_\Gamma\}$ as $h_\Gamma \to 0$ have the same sense as before. Besides, the fundamental sequences $\{u_\Gamma\}$ and $\{v_\Gamma\}$ are equivalent if the sequence of difference $\{u_\Gamma - v_\Gamma\}$ tends to zero. Particularly, we had the fundamentality with respect to the weak topology of the space $H_0^1(0, L)$ for the considered example.

Definition 9.2 *If the sequence $\{u_\Gamma\}$ is fundamental with respect to a linear topological space, then this is called the **sequential model**, and the corresponding equivalence class is called the **sequential state**.*

Remark 9.14 Of course, these states and models depend upon the functional space, where we determine the fundamentality of the sequence of the approximate states.

TABLE 9.4: Process of construction of the mathematical models

object	general case	heat transfer
set	Ω	$[0, L]$
elementary volume	Ω_α	$\Omega_i = (x_{i-1}, x_i)$
partition of the set	$\Gamma = \{\Omega_\alpha\}$	$\Gamma_h = \{(x_{i-1}, x_i)\}$
discrete state	$\bar{u}_\Gamma = \{u_\alpha\}$	$\bar{u}_\Gamma = \{u_i\}$
discrete model	$A_\alpha \bar{u}_\Gamma = 0 \ \forall \alpha$	$\delta_x(k_i \delta_{\bar{x}} u_i) = f_i, \ i = 1, ..., M-1$ $u_0 = 0, \ u_M = 0$
space of grid functions	V_Γ	V_h
prolongation of discrete state	$u_\Gamma = \pi \bar{u}_\Gamma$	linear interpolation u_h of the discrete state
space of fundamentality	H	space $H_0^1(0, L)$ with weak topology
sequential model	$\{u_\Gamma\}$	$\{u_h\}$
sequential state	$\lfloor u_\Gamma \rfloor$	$\lfloor u_h \rfloor$
space of convergence	U	space $H_0^1(0, L)$ with weak topology
space of test functions	Λ	$H_0^1(0, L)$
generalized model	$B(u, \lambda) = 0 \ \forall \lambda \in \Lambda$	$\int\limits_0^L -k\frac{d\lambda}{dx}\frac{du}{dx}dx = \int\limits_0^L \lambda f dx \ \ \forall \lambda \in H_0^1(0, L)$
space of regular functions	V	$C_0^2[0, L]$
classic model	$Au = 0$	$[k(x)u'(x)]' = f(x), \ x \in (0, L)$ $u(0) = 0, \ u(L) = 0$

Obviously, Definition 9.1 is a partial case of Definition 9.2. If any real or p-adic number is approximated by rational numbers, the arbitrary distribution by infinitely differentiable functions, and the sequential control by usual controls, then the sequential state of the system is approximated by its approximate states, which are determined by the relations (9.19) for each partition of the given set. When the domain Ω is unboundedly divided into cells, the set of approximate states can be fundamental in the indicated sense. Then it determines a sequential model of the system and thereby determines its sequential state.

It is necessary to analyze the system of approximate solutions only for checking the fundamentality of the system. This information is known. We do not use the solution of a problem that we initially do not know anything about. An infinite number of sequential models corresponds to a specific sequential model, just as a real number can be approximated by different rational sequences, a generalized function by different sequences of regular functions, and sequential control by many sequences of usual controls. This circumstance gives a certain freedom of choice of a specific algorithm in the practical solution of the problem.

Determine the connection between the introduced concepts and the results obtained on the basis of the classical and generalized approaches. Let $(\cdot, \cdot)_\Gamma$ be a scalar product on the set V_Γ of the grid functions. Then the equality (9.19) is equivalent to

$$\left(A_\Gamma \bar{u}_\Gamma, \bar{\lambda}_\Gamma\right)_\Gamma = 0 \ \ \forall \bar{\lambda}_\Gamma \in V_\Gamma. \tag{9.20}$$

This is the equality (9.15) for our example.

Consider linear topological spaces U and Λ of functions that are determined on the set Ω such that U is a subspace of H. Suppose there exists a functional B_Γ on the product of U and Λ that is linear with respect to the second argument such that

$$\left(A_\Gamma \bar{u}_\Gamma, \bar{\lambda}_\Gamma\right)_\Gamma = B_\Gamma\left(\pi \bar{u}_\Gamma, \pi_\lambda \bar{\lambda}_\Gamma\right) \forall \bar{u}_\Gamma, \bar{\lambda}_\Gamma \in V_\Gamma,$$

where π_λ is a prolongation operator from V_Γ to Λ. Particularly, for the stationary heat transfer phenomenon U is the set $H_0^1(0, L)$ with weak topology, Λ is the same set with strong topology, π_λ is the linear interpolation of the grid function, i.e. $\pi_\lambda = \pi$, and the operator B_Γ is determined by the formula

$$B_\Gamma(u, \lambda) = \int_0^L k^h \frac{d\lambda}{dx} \frac{du}{dx} dx + \int_0^L g^h \frac{d\lambda}{dx} dx \quad \forall u \in U, \lambda \in \Lambda.$$

Now the equality (9.20) is transformed to

$$B_\Gamma\left(\pi \bar{u}_\Gamma, \pi_\lambda \bar{\lambda}_\Gamma\right) = 0 \ \forall \bar{\lambda}_\Gamma \in V_\Gamma \tag{9.21}$$

that is the analogue of (9.16). Suppose the set of all prolongations $\{\pi_\lambda \bar{\lambda}_\Gamma\}$ is dense in Λ. Then for any $\lambda \in \Lambda$ we can choose the values $\bar{\lambda}_\Gamma$ such that

$$\pi_\lambda \bar{\lambda}_\Gamma \to \lambda \ \text{in} \ \Lambda \tag{9.22}$$

as $h_\Gamma \to 0$. Suppose also the following convergence

$$\pi \bar{u}_\Gamma \to u \ \text{in} \ U. \tag{9.23}$$

The conditions (9.22), (9.23) for the considered example are true by the proof of the Theorem 9.2.

Suppose now there exists an operator B on the product $\Lambda \times U$ that is linear with respect to the first argument such that from the conditions (9.22), (9.23) the convergence follows

$$B_\Gamma\left(\pi \bar{u}_\Gamma, \pi_\lambda \bar{\lambda}_\Gamma\right) \to B(u, \lambda). \tag{9.24}$$

For our case the functional B is determined by the formula

$$B(u, \lambda) = -\int_0^L k \frac{d\lambda}{dx} \frac{du}{dx} dx - \int_0^L f\lambda dx \quad \forall u, \lambda \in H_0^1(0, L).$$

After passing to the limit at the equality (9.21) with using the condition (9.24) we get

$$B(u, \lambda) = 0 \ \forall \lambda \in \Lambda \tag{9.25}$$

that is an analogue of the equality (9.15). Now we can give the definition of the generalized model of the system that has the generalized model of the stationary heat transfer phenomenon as the partial case.

Definition 9.3 *The **generalized model** of the system is the equality (9.25), and its **generalized state** is the function $u \in U$ that satisfies this equality.*

By this definition, the sequence $\{u_\Gamma\}$ is convergent, and therefore, fundamental in the sense of the space U. Besides, its limit u coincides up to isomorphism with the class of equivalent sequences with the representation $\{u_\Gamma\}$. Similarly, the limit of a convergent fundamental sequence of rational numbers (respectively, smooth functions and usual controls) is identified with a regular equivalence class defined by a given sequence with a real number (respectively, a distribution and sequential control). Thus, the generalized state of the system necessarily turns out to be its sequential state.

Definition 9.3 applies stronger requirements to the sequential model than Definition 9.2. Therefore, the existence of a sequential state of the system is possible in the absence of a generalized state. In this case, the sequential state can be called a **sequential solution** of problem (9.25). It is clear that under these conditions this relation is only a formal expression in the same way as in the absence of a classical solution the boundary value problem itself (in the usual interpretation) does not make sense, but is understood only as a short notation of the corresponding integral equality.

Now consider the classic approach. Let V be a linear topological space such that embedding V to U is continuous. Suppose there exists an operator A that is determined on the space V such that the following equality holds

$$\langle Au, \lambda \rangle = (u, \lambda) \ \forall u \in U, \lambda \in \Lambda.$$

For our example V is the space $C_0^2[0, L]$ of twice continuously differentiable functions on the interval $[0, L]$ with zero value at the ends of the interval, and the operator A is determined by the formula

$$Au = \frac{d}{dx}\left(k\frac{du}{dx}\right) - f.$$

From the equality (9.25) it follows that

$$\langle Au, \lambda \rangle = 0 \ \forall \lambda \in \Lambda.$$

Then we obtain

$$Au = 0. \tag{9.26}$$

Definition 9.4 *The **classic model** of the system is the equality (9.26), and its **classic state** is the function $u \in U$ that satisfies this equality.*

If there exists a classical state of the system, then (after choosing of functional spaces), this is a generalized state of the system, and hence its sequential state. However, in Definition 9.4, even stronger requirements are imposed on the sequential model $\{u_\Gamma\}$. Then a situation is possible where the classical state of the system is absent, but there is only its generalized or even only its sequential state (see Figures 9.5 and 9.6). In this case, a generalized (sequential) state can be called a generalized (respectively, sequential) solution of equation (9.21), in contrast to its classical solution, which is a classical state of the system. In this case, equation (9.26) should be understood only as a formal abbreviated record of the corresponding generalized or sequential model. The obtained results give a certain hope for the possibility of a wide application of the sequential approach in mathematical physics.

9.6   Conclusions

1. By physical laws, the balance relations that describe the processes occurring in the elementary volume of the object under consideration are determined.

2. The classical method of determination of mathematical physics equations uses the passage to the limit in the balance relations that requires the a priori properties of the state function of the system. One cannot possibly check it before the determination of the mathematical model.

3. An attempt to determine a generalized model of the system, connected with the passage to the limit after multiplying the existing equality by some test function and integrating over a given set, leads to analogous difficulties.

4. To overcome the difficulties that have arisen, a sequential method is used that is related to the Cauchy convergence criterion and the completion of a metric space. This provides the possibility of justifying the convergence of the sequences.

5. The sequential method is successfully used for the constructive determination of the real and the p-adic numbers, the sequential controls and the distributions.

6. The application of the sequential approach to the problems of mathematical physics begins with the construction of a discrete model of the system on the basis of available balance relationships and does not require any mathematical assumptions.

7. By the discrete model for an arbitrary partition of the domain into elementary parts, the corresponding grid function and its interpolation to the whole given set can be found, which is interpreted as the "approximate state" of the system.

8. If the sequence of "approximate solutions" is fundamental, then it is called a sequential model of the system, and the equivalent equivalence class corresponding to it is the sequential state of the system.

9. The justification of the sequential approach does not require any a priori suppositions with respect to the state function.

10. By the sequential approach, the determination of the model, the proof of the existence of the system state and the justification of the algorithm for approximate solution of the problem are realized simultaneously.

11. The approximate value of the sequential state of the system is determined from the discrete model underlying the practical finding of the classical and generalized state of the system.

12. The approximate value of the sequential state of the system is the linear interpolation of the solution of the discrete model.

13. Under an additional supposition, the sequence of the approximate solutions tends toward the generalized state of the system.

14. The generalized state of the system is unique.

15. Each generalized state of the system is determined by its sequential state. However, not every sequential state is reduced to a generalized one.

16. The classic state of the system is unique.

17. The practical solving of the discrete model can give the sequential, generalized, and classic state. The result depends upon the convergence properties.

18. The sequential method not only determines the weaker form of the model, which is realized even in cases when it is not possible to obtain other forms of models, but also gives a justification for the process of constructing the classical and generalized models.

This problem is interesting not only in itself. In the process of its research, we encountered various seemingly unrelated mathematical directions. This is an excellent illustration of the fact that mathematics is a unified science, and the boundaries of its directions are very conditional.

Notes

[1] The methods of solving the boundary problems for difference equations is described, for example in [15], [41], [130], [139], [141], [150]. [152], [180].

[2] The considered definition is given in [166], [169].

[3] The finite difference method is described, for example in [15], [41], [130], [139], [141], [150]. [152], [180].

[4] The discrete spaces and their properties are described, for example, in [14], [114].

[5] The general theory of linear continuous functionals and the Riesz theorem are described by standard courses of functional analysis (see, for example, [42], [71], [81], [91], [95], [100], [142], [147].)

[6] The Theorem 9.1 is considered in [166].

[7] One can prove the existence of the general solution of the considered boundary problem by means of the Galerkin method, the variational method, and the Ritz method; see, for example, [53], [101], [103], [116], [114], [115], [126], [128], [195].

[8] The boundary problems for the second order differential equations are considered in [36], [65], [80], [111], [138]. Consider, for example, the problem

$$u''(x) + au(x) = 1, \ 0 < x < \pi, \ u(0) = 0, \ u(\pi) = 0,$$

where a is a constant. This differential equation has the general solution

$$u(x) = c_1 \sin \sqrt{a}x + c_2 \cos \sqrt{a}x + 1/a,$$

where c_1 and c2 are arbitrary constants. These constants can be found from the given boundary conditions. For example, for $a = 2$ we find its values and obtain the unique solution of the boundary problem by the previous formula. However, for $a = 4$ both boundary conditions are transformed to the same equality

$$c_2 + 1/4 = 0.$$

Therefore, we can find here c_2, but the constant c1 will be arbitrary. Thus, the boundary problem has an infinite set of solutions. Finally, for $a = 1$, from boundary conditions we get the equalities

$$c_2 + 1 = 0, \ -c_2 + 1 = 0.$$

We have the contradiction, and the boundary problem is insolvable. These properties are related to the **spectral theory** for operators; see, for example, in [42], [71], [73], [97].

[9]Consider the nonlinear **Chafee–Infante boundary problem**, proposed by *Nathaniel Chafee* and *Ettore Infante* [34] (see also [66])

$$u''(x) + au(x) - b[u(x)]^3 = 0, \ 0 < x < \pi, \ \ u(0) = 0, \ u(\pi) = 0,$$

where a and b are positive constants. One can prove (see [34]) that under the inequality $(k-1)^2 < b < k^2$, where k is an arbitrary natural number, the boundary problem has $2k - 1$ solutions

[10]The filter theory is described in [26], and the net theory is described in [83].

Bibliography

[1] R.A. Adams and J.J.F. Fournier. *Sobolev Spaces*. Academic Press, 2003.

[2] I.T. Adamson. *Introduction to Field Theory*. Dover Publications, 2007.

[3] R.P. Agarwal, M. Meehan, and D. O'Regan. *Fixed Point Theory and Applications*. Cambridge University Press, 2001.

[4] S. Agmon. *Lectures on Elliptic Boundary Value Problems*. Amer. Math. Soc., Chelsea Publications, 2010.

[5] A.K. Aksoy and M.A. Khamsi. *Nonstandard Methods in Fixed Point Theory*. Springer Verlag, 1990.

[6] P.S. Aleksandrov. *Introduction to the Sets Theory and the General Topology*. Nauka, Moskow, 1977.

[7] P.S. Aleksandrov and B.A. Pasynkov. *Introduction to the Theory of Dimension. Introduction to the Theory of Topological Spaces and General Theory of Dimension*. Nauka, Moscow, 1973.

[8] V.M. Alekseev, V.M. Tihomirov, and S.V. Fomin. *Optimal Control*. Nauka, Moscow, 1979.

[9] R. Allenby. *Rings, Fields and Groups*. Butterworth-Heinemann, 1991.

[10] P.V. Ananda Mohan. *Residue Number Systems*. Springer International Publishing, Cham, 2016.

[11] R. Antosik, J. Mikusinski, and R. Sikorski. *Theory of Distributions. The Sequential Approach*. Elsevier, 1973.

[12] T. Apostol. *Mathematical Analysis*. Addison-Wesley, 1974.

[13] V.I. Arnold. *Ordinary Differential Equations*. Springer-Verlag, Berlin, Heidelberg, 1992.

[14] J.P. Aubin. *Approximation of Elliptic Boundary-Value Problems*. Wiley-Interscience, 1972.

[15] N.S. Bahvalov. *Numerical Methods*. Nauka, Moscow, 1975.

[16] R. Baierlein. *Thermal Physics*. Cambridge University Press, 1999.

[17] M. Bailyn. *A Survey of Thermodynamics*. American Institute of Physics Press, New York, 1994.

[18] J. Baranger. *Quelque resultats en optimization non convexe*. Thèse, Grenoble, 1973.

[19] E.A. Bender. *An Introduction to Mathematical Modeling*. Dover, New York, 2000.

[20] L. Bers, F. John, and M. Schechter. *Partial Differential Equations*. American Mathematical Soc., 1964.

[21] O.V. Besov, V.P. Il'in, and S.M. Nikol'ski. *Integral Representations of Functions and Embedding Theorems*. Nauka, Moscow, 1975.

[22] G. Bliss. *Calculus of Variations*. Open Court, 1944.

[23] Z.I. Borevich and I.R. Shafarevich. *Theory of Numbers*. Nauka, Moscow, 1985.

[24] N. Bourbaki. *Algebra II*. Springer, 1988.

[25] N. Bourbaki. *Algebra I*. Springer, 1989.

[26] N. Bourbaki. *General Topology*. Springer, 1989.

[27] N. Bourbaki. *Theory of Sets*. Springer Science and Business Media, 2004.

[28] N. Bourbaki. *Espaces Vectoriels Topologiques*. Springer, 2007.

[29] H. Bremermann. *Distributions, Complex Variables and Fourier Transforms*. Addison-Wesley Publishing, 1965.

[30] V. Bryant. *Metric Spaces: Iteration and Application*. Cambridge University Press, 1985.

[31] I. Bucur and Deleanu A. *Introduction to the Theory of Categories and Functors*. Wiley, 1968.

[32] H.S. Carslaw and J.C. Jaeger. *Conduction of Heat in Solids*. Clarendon Press, Oxford, 1959.

[33] L. Cesari. Existence theorems for problems of weak and usual optimal solutions in Lagrange problems with unilateral constraints. *Trans. Amer. Math. Soc.*, 124(3):369–430, 1966.

[34] N. Chafee and E.F. Infante. Bifurcation and stability for a nonlinear parabolic partial differential equation. *Bull. Amer. Math. Soc.*, 80(1):49–52, 1974.

[35] M. Chester. *Primer of Quantum Mechanics*. John Wiley, 1987.

[36] E. Coddington and N. Levinson. *Theory of Ordinary Differential Equations.* McGraw-Hill, New York, 1955.

[37] P.J. Collins and A.W. Roscoe. Criteria for Metrisability. *Proceedings of the American Mathematical Society,* 90(4):631–640, 1984.

[38] J.F. Colombeau. *New Generalized Functions and Multiplication of Distributions.* Amsterdam, North Holland, 1984.

[39] J.F. Colombeau. *Elementary Introduction to New Generalized Functions.* Amsterdam, North Holland, 1985.

[40] M. Comenetz. *Calculus: The Elements.* World Scientific, 2002.

[41] Biswa Nath Datta. *Numerical Linear Algebra and Applications.* SIAM, 2010.

[42] J. Diestel. *Geometry of Banach Spaces.* Springer-Verlag, Berlin, Heidelberg, 1975.

[43] Yu.N. Drozhzhinov and B.I. Zavyalov. *Introduction to the Theory of Generalized Functions.* MIAN, Moscow, 2006.

[44] G. Dubois. *The Nature of Mathematical Modeling.* Cambridge University Press, 1998.

[45] W. Dunham. *The Calculus Gallery.* Princeton University Press, 2005.

[46] G. Duvaut and J.L. Lions. *Inequalities in Mechanics and Physics.* Springer Verlag, Berlin, Heidelberg, New York, 1976.

[47] Yu.V. Egorov. On the Theory of Generalized Functions. *Uspehi math. nauk,* 45(5):3–40, 1990.

[48] E.H. Moore and H.L. Smith. A general theory of limit. *Amer. J. Math.,* 44:102–121, 1922.

[49] I. Ekeland and R. Témam. *Convex Analysis and Variational Problems.* North-Holland, Amsterdam, 1976.

[50] L.E. Elsgolc. *Calculus of Variations.* Pergamon Press Ltd, 1962.

[51] R. Engelking. *General Topology.* PWN, 1977.

[52] A. Faghri, Y. Zhang, and J. Howell. *Advanced Heat and Mass Transfer.* Columbia, MO: Global Digital Press, 2010.

[53] S. Farlow. *Partial Differential Equations for Scientists and Engineers.* John Wiley and Sons, New York, 1982.

[54] W.H. Fleming and R.W. Rishel. *Deterministic and Stochastic Optimal Control.* Springer, New York, 1975.

[55] A.A. Frenkel, Y. Bar-Hillel, and A. Levy. *Foundation of Set Theory.* Elsevier, 1973.

[56] R. Gabasov and F.M. Kirillova. *Special Optimal Controls.* Nauka, Moscow, 1973.

[57] B. Gelbaum and J. Olmsted. *Counterexamples in Analysis.* Holden-Day, Inc., San Francisco, 1962.

[58] I.M. Gelfand and S.V. Fomin. *Calculus of Variations.* Courier Corporation, 2000.

[59] N. Gershenfeld. *Modeling and Simulation.* Taylor and Francis, CRC Press, 2018.

[60] R. Glowinski, J.-L. Lions, and R. Tremolieres. *Numerical Analysis of Variational Inequalities.* North-Holland, Amsterdam, 1981.

[61] S.K. Godunov and V.S. Ryaben'ky. *Difference Schemes. Introduction to Theory.* Nauka, Moscow, 1977.

[62] J. Hadamard. *Le problème de Cauchy et les Equations aux Derivées Partielles Lineaires Hyperbolic.* Hermann, Paris, 1932.

[63] P. Halmos. *Finite-Dimensional Vector Spaces.* Springer, 1958.

[64] P. Halmos. *Measure Theory.* Springer, 1964.

[65] P. Hartman. *Ordinary Differential Equations.* SIAM, 2002.

[66] D. Henry. *Geometric Theory of Semilinear Parabolic Equations.* Springer-Verlag, 1981.

[67] H. Herrlich and G. Strecker. *Category Theory.* Heldermann Verlag Berlin, 2007.

[68] A. Heyting. *Intuitionism: an Introduction.* North-Holland Pub. Co., 1971.

[69] Y. Hirata and H. Ogata. On the exchange formula for distributions. *J. Sci. Hiroshima Univ. Ser.A-I*, 22(3):147–152, 1958.

[70] J.M. Howie. *Real Analysis.* Springer, 2005.

[71] V. Hutson, J.S. Pym, and Cloud M.J. *Applications of Functional Analysis and Operator Theory.* Elsevier Science, 2005.

[72] F.P. Incropera et al. *Fundamentals of Heat and Mass Transfer.* Wiley, 2012.

[73] A.D. Ioffe and V.M. Tihomirov. *Theory of Extremal Problems.* Nauka, Moscow, 1974.

[74] A. Iserles. *A First Course in the Numerical Analysis of Differential Equations*. Cambridge University Press, 1996.

[75] M. Itano. On the multiplicative products of distributions. *J. Sci. Hiroshima Univ. Ser.A-I*, 29(1):51–74, 1965.

[76] V.K. Ivanov. Hyperdistributions and the multiplication of Schwartz distributions. *DAN USSR*, 204(5):1045–1048, 1972.

[77] T. Jech. *Set Theory*. Springer-Verlag, Berlin, New York, 2003.

[78] J. Jost and X. Li-Jost. *Calculus of Variations*. Cambridge University Press, 1998.

[79] D. Kahaner, L. Moler, S. Nash, and Forsythe G. *Numerical Methods and Software*. Prentice Hall, 1989.

[80] E. Kamke. *Differentialgleichungen Reeller Funktionen*. Akademische Verlagsgesellschaft, Leipzig, 1962.

[81] S. Kantorovitz. *Introduction to Modern Analysis*. Oxford University Press, 2006.

[82] M.I. Kargapolov and Yu.I. Merzlyakov. *Foundations of the Group Theory*. Nauka, Moscow, 1972.

[83] J. Kelley. *General Topology*. Springer, 1955.

[84] D. Kinderlehrer and G. Stampacchia. *An Introduction to Variational Inequalities and Their Applications*. Academic Press, Boston, 1980.

[85] A.A. Kirillov. *What is the Number?* Nauka, Moscow, 1993.

[86] D.E. Kirk. *Optimal Control Theory: An Introduction*. New Jersey, Englewood Cliffs, 2004.

[87] W.A. Kirk and K. Goebel. *Topics in Metric Fixed Point Theory*. Cambridge University Press, 1990.

[88] W.A. Kirk and M.A. Khamsi. *An Introduction to Metric Spaces and Fixed Point Theory*. John Wiley, New York, 2001.

[89] N. Koblitz. *p-adic Numbers, p-adic Analysis, and Zeta-Functions*. Springer-Verlag, New York, 1984.

[90] A.N. Kolmogorov. To the substantiation of the theory of real numbers. *Uspehi Math. Nauk*, 1(1):217–219, 1946.

[91] A.N Kolmogorov and S.V. Fomin. *Elements of the Theory of Functions and Functional Analysis*. Dover Publications, 1999.

[92] H. König. Multiplikations und Variablentransformation in der Theorie der Distributionen. *Arch. Math.*, (6):391–396, 1956.

[93] G. Köthe. *Topological Vector Spaces.* Springer-Verlag, Berlin, New York, 1969.

[94] A. Krasnosel'skii and P.P. Zabreiko. *Geometric Methods of Nonlinear Analysis.* Springer-Verlag, Berlin, Heidelberg, 1984.

[95] S.G. Krein et al. *Functional Analysis.* Nauka, Moscow, 1964.

[96] R. Kress. *Numerical Analysis Springer.* Princeton University Press, 1998.

[97] V.F. Krotov. *Global Methods in Optimal Control Theory.* Marcel Dekker, New York, 1996.

[98] K. Kuratowski. *General Topology.* Academic Press, 2014.

[99] K. Kuratowski and A. Mostowski. *Set Theory.* Polish Scientific Publishers, 1968.

[100] S.S. Kutateladze. *Fundamentals of Functional Analysis.* Kluwer Academic Publishers, Dordrecht, 1995.

[101] O.A. Ladyzhenskaya. *The Boundary Value Problems of Mathematical Physics.* Springer Verlag, Berlin-Heidelberg-New York, 1985.

[102] O.A. Ladyzhenskaya, V.A. Solonnikov, and N.N. Ural'ceva. *Linear and Quasi-linear Equations of Parabolic Type.* Providence, RI: American Mathematical Society, 1968.

[103] O.A. Ladyzhenskaya and N.N. Ural'ceva. *Linear and Quasilinear Elliptic Equations.* Academic Press, New York and London, 1968.

[104] T.Y. Lam. *Lectures on Modules and Rings.* Springer-Verlag, New York, 1999.

[105] T.Y. Lam. *Exercises in Classical Ring Theory.* Springer-Verlag, New York, 2003.

[106] J.D. Lambert. *Numerical Methods for Ordinary Differential Systems.* John Wiley and Sons, Chichester, 1991.

[107] L.D. Landau and E.M. Lifshitz. *Quantum Mechanics: Non-Relativistic Theory.* Pergamon Press, 1977.

[108] S. Lang. *Algebra.* Springer-Verlag, New York, 2002.

[109] R. Larson and B. Edwards. *Calculus of a Single Variable.* Cengage Learning, 2010.

[110] D. Lay. *Linear Algebra and Its Applications*. Pearson, 2012.

[111] S. Lefschetz. *Differential Equations: Geometric Theory*. Interscience Publishers, 1963.

[112] T. Leinster. *Basic Category Theory*. Cambridge University Press, 2014.

[113] F.L. Lewis. *Optimal Control*. John Wiley and Son, New York, 1986.

[114] J.-L. Lions. *Quelques méthodes de résolution des problèmes aux limites non linéaires*. Dunod, Paris, 1969.

[115] J.-L. Lions and E. Magenes. *Problemes aux Limites non Homogenes et Applications*. Dunod, Paris, 1968.

[116] J.L. Lions. *Contrôle optimal de systèmes gouvernés par des équations aux dérivées partielles*. Dunod, Gauthier-Villars, Paris, 1968.

[117] C. Long. *Elementary Introduction to Number Theory*. Lexington, 1972.

[118] A.V. Lykov. *Theory of Heat Conductivity*. VSh, Moscow, 1967.

[119] G. Mackey. *The Mathematical Foundations of Quantum Mechanics*. Dover Publications, 2004.

[120] Yu.I. Manin and A.A. Panchishkin. *Introduction to the Numbers Theory*. VINITI, Moscow, 1990.

[121] P. Martin-Löf. *Constructive Mathematics and Computer Programming*. North-Holland, Amsterdam, 1979.

[122] Yu.V. Matiyasevich. *10-th Hilbert Problem*. Nauka, Moscow, 1993.

[123] K. Maurin. *Methods of Hilbert Spaces*. Polish Scientific Publishers, 1967.

[124] V. Maz'ja. *Sobolev Spaces*. Springer-Verlag, Berlin, Heidelberg, New York, 1985.

[125] E.J. McShane. Generalized curves. *Duke Math. J.*, (6):513–536, 1940.

[126] S.G. Mikhlin. *Linear Equations of Mathematical Physics*. Holt, Rinehart and Winston, 1967.

[127] J. Mikusinski. Une definition de distribution. *Bull. Acad. Polon. Sci. Cl.1*, 3(11):589–591, 1955.

[128] S. Mizohata. *The Theory of Partial Differential Equations*. Cambridge University Press, Cambridge, 1979.

[129] L.J. Mordell. *Diophantine Equations*. Academic Press, 1969.

[130] K.W. Morton and D.F. Mayers. *Numerical Solution of Partial Differential Equations, An Introduction*. Cambridge University Press, 2005.

[131] F. Murat. Thèoremes de non existènce pour des problèmes de contrôle dans les coefficients. *C. R. Acad. Sci. Paris*, (5):A395–A398, 1972.

[132] K. Nicholson. *Introduction to Abstract Algebra*. John Wiley and Sons, 2012.

[133] S.M. Nikol'ski. *Approximation of Functions of Many Variables and Embedding Theorems*. Nauka, Moscow, 1977.

[134] D. Olander. *General Thermodynamics*. CRC Press, 2007.

[135] A. Papadopoulos. *Metric Spaces, Convexity and Nonpositive Curvature*. European Mathematical Society, 2004.

[136] Ph.E. Gill, W. Murray, and Wright M.H. *Practical Optimization*. Academic Press, London, 1981.

[137] Y. Pinchover and J. Rubinstein. *An Introduction to Partial Differential Equations*. Cambridge University Press, Cambridge, 2005.

[138] A.D. Polyanin and V.F. Zaitsev. *Handbook of Exact Solutions for Ordinary Differential Equations*. Chapman and Hall/CRC Press, Boca Raton, 2003.

[139] A. Quarteroni, R. Sacco, and F. Saleri. *Numerical Mathematics*. Springer, New York, 2007.

[140] R. Fletcher. *Practical Optimization Methods*. John Wiley and Son, Chichester, 1987.

[141] J. Randall. *Finite Difference Methods for Ordinary and Partial Differential Equations*. SIAM, 2007.

[142] M. Reed and B. Simon. *Functional Analysis*. Academic Press, 1980.

[143] R.D. Richtmyer and K.W. Morton. *Difference Methods for Initial-Value Problems*. Interscience Publisher, New York, 1967.

[144] A.P. Robertson and W.J. Robertson. *Topological Vector Spaces*. Cambridge Tracts in Mathematics, 1964.

[145] M. Ronan. *Symmetry and the Monster*. Oxford University Press, 2006.

[146] J. Rotman. *An Introduction to the Theory of Groups*. Springer-Verlag, New York, 1994.

[147] W. Rudin. *Functional Analysis*. McGrawHill, 1991.

[148] Yu.B. Rumer and M.Sh. Ryvkin. *Thermodynamics, Statistical Physics, and Kinetics*. MIR Publishers, 1980.

[149] H. Sagan. *Introduction to the Calculus of Variations*. Dover, 1992.

[150] A.A. Samarsky. *Theory of Difference Schemes.* Nauka, Moscow, 1989.

[151] A.A. Samarsky and Gulin A.V. *Numerical Methods.* Nauka, Moscow, 1989.

[152] A.A. Samarsky and E.S. Nikolaev. *Methods of Solving of Grid Equations.* Nauka, Moscow, 1978.

[153] H. Schaefer and M.P. Wolff. *Topological Vector Spaces.* Springer Verlag, 1999.

[154] P. Schapira. *Theories des Hyperfonctions.* Springer, 1970.

[155] W. Schmidt. *Diophantine Approximations and Diophantine Equations.* Springer-Verlag, Berlin, 1991.

[156] B. Schröder. *Ordered Sets: An Introduction.* Birkhäuser, Boston, 2003.

[157] L. Schwartz. Sur l'Iimpossibilité de la Multiplication des Distributions. *C. r. Acad. Sci. Paris,* 239(15):847–848, 1954.

[158] L. Schwartz. *Théorie des Distributions.* Hermann, 1966.

[159] W.R. Scott. *Group Theory.* Dover, New York, 1987.

[160] S. Serovaisky. *Counterexamples in the Optimal Control Theory.* Brill Academic Press. Netherlands, Utrecht–Boston, 2004.

[161] S. Serovajsky. *Mathematical Modeling.* Kazak University, Almaty, 2000.

[162] S. Serovajsky. On the effectiveness of the topological multiplicative theory of distributions. *Izv. NAN RK, ser. phys.-math.,* (5), 2000.

[163] S. Serovajsky. Lower completion and extension of the extremum problems. *Izv. vuzov. Mathematics,* (5):30–41, 2003.

[164] S. Serovajsky. Sequential extension in the coefficients control problems for elliptic type equations. *J. of Inverse and Ill-Posed Problems,* 11(5):523–536, 2003.

[165] S. Serovajsky. Sequential optimal control. *Izv. NAN RK, phys.-math. ser.,* (1):50–56, 2003.

[166] S. Serovajsky. *Sequential Models of Mathematical Physics.* Print-S, Almaty, 2004.

[167] S. Serovajsky. Weak approximate solution of the optimization problem for a nonlinear elliptic equation without convexity of the admissible control set. *Review Bulletin of the Calcutta Mathematical Society,* 20(1):51–58, 2012.

[168] S. Serovajsky. Optimal control for systems described by hyperbolic equation with strong nonlinearity. *Journal of Applied Analysis and Computation*, 3(2):183–195, 2013.

[169] S. Serovajsky. Sequential models of physical phenomenon and justification of mathematical modeling. *Advanced Mathematical Models and Applications*, 2(1):6–13, 2017.

[170] S. Serovajsky. State-constrained optimal control of nonlinear elliptic variational inequalities. *Optimization Letters*, 8(7):2041–2051, 2017.

[171] J.-P. Serre. *A Course in Arithmetic*. Springer, 1996.

[172] B.V. Shabat. *Introduction to Complex Analysis*. Nauka, Moscow, 1969.

[173] G. Simmons. *Introduction to Topology and Modern Analysis*. McGraw-Hill, New York, 1963.

[174] L.A. Skornyakov et al. *General Algebra. Vol. 1*. Nauka, Moscow, 1990.

[175] L.A. Skornyakov et al. *General Algebra. Vol. 2*. Nauka, Moscow, 1991.

[176] G.D. Smith. *Numerical Solution of Partial Differential Equations: Finite Difference Methods*. Oxford University Press, 1985.

[177] S.L. Sobolev. *Some Applications of Functional Analysis in Mathematical Physics*. American Math. Soc., 2008.

[178] J. Stoer and R. Bulirsch. *Introduction to Numerical Analysis*. Springer-Verlag, New York, 1980.

[179] G. Strang. *Linear Algebra and Its Applications*. Thomson Learning, Inc., 2006.

[180] J. Strikwerda. *Finite Difference Schemes and Partial Differential Equations*. SIAM, 2004.

[181] M.B. Suryanarayana. Existence theorems for problems concerning hyperbolic partial equations. *J. Optimal Theory Appl.*, 15(4):361–392, 1975.

[182] S.A. Telyakovsky. *Course of Lectures on Mathematical Analysis*. MIAN, Moscow, 2009.

[183] G. Teschl. *Ordinary Differential Equations and Dynamical Systems*. American Mathematical Society, Providence, 2012.

[184] A.N. Tihonov and Samarsky A.A. *Equations of Mathematical Physics*. Nauka, Moscow, 1977.

[185] A.N. Tihonov and V.Ya. Arsenin. *Methods of Solving of Ill-posed Problems*. Nauka, Moscow, 1979.

[186] H.G. Tillmann. Darstellung der Shwartzchen Distriburionen Durch Analytische Funktionen. *Math. Z.*, 77(2):106–124, 1961.

[187] E.C. Titchmarsh. *The Theory of Functions.* Oxford University Press, 1939.

[188] G. Tolstov. *Fourier Series.* Courier-Dover, 1976.

[189] W.F. Trench. *Introduction to Real Analysis.* Pearson Education, 2013.

[190] P.N. Vabishchevich et al. *Computational Methods in the Mathematical Physics.* Moscow University, Moscow, 1986.

[191] B. van der Waerden. *Algebra.* Springer-Verlag, 1960.

[192] F.P. Vasiliev. *Methods of Solving Extremum Problems.* Nauka, Moscow, 1981.

[193] V.S. Vladimiriv and I.V. Volovich. *P*-adic quantum mechanics. *Commun. Math. Phys.*, 123:659–676, 1989.

[194] V.S. Vladimiriv and I.V. Volovich. *P*-adic Schrodinger-type equation. *Lett. Math. Phys.*, 18:43–53, 1989.

[195] V.S. Vladimirov. *Equations of Mathematical Physics.* M. Dekker, 1971.

[196] V.S. Vladimirov. *Generalized Functions in Mathematical Physics.* Taylor and Francis, 2002.

[197] I.V. Volovich. *P*-adic string. *Class. Quant. Grav.*, 4:L83–L84, 1987.

[198] J.P. Ward. *Quaternions and Cayley Numbers: Algebra and Applications.* Kluwer Academic Publishers, 1997.

[199] J. Warga. *Optimal Control of Differential and Functional Equations.* Academic Press, 1972.

[200] C. Whitehead. *Guide to Abstract Algebra.* Palgrave, Houndmills, 2002.

[201] N. Wiener. *Cybernetics: or the Control and Communication in the Animal and the Machine.* MIT Press, 1961.

[202] I.M. Yaglom. *Mathematical Structures and Mathematical Modelling.* Gordon and Breach Science Publishers Ltd, 1984.

[203] K. Yosida. *Functional Analysis.* Springer, 1995.

[204] L. Young. *Lectures on the Calculus of Variations and Optimal Control Theory.* W.B. Saunders Co., Philadelphia–London–Toronto, 1969.

[205] T. Zolezzi. A characterization of well-posed optimal control systems. *SIAM J. Control and Optim.*, 19(5):604–616, 1981.

[206] D. Zwillinger. *Handbook of Differential Equations.* Academic Press, Boston, 1997.

[207] A. Zygmund. *Trigonometric Series.* Cambridge University Press, Cambridge, 2002.

Index

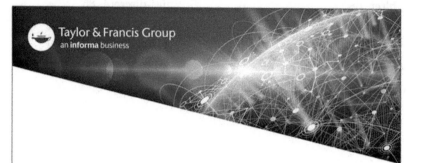